Graduate Texts in Mathematics **117**

Springer
New York
Berlin
Heidelberg
Barcelona
Budapest
Hong Kong
London
Milan
Paris
Tokyo

Graduate Texts in Mathematics

continued after index

Jean-Pierre Serre

Algebraic Groups and Class Fields

Translation of the French Edition

 Springer

Jean-Pierre Serre
Professor of Algebra and Geometry
Collège de France
75231 Paris Cedex 05
France

AMS Classifications: 11G45 11R37

LCCN 87-32121

This book is a translation of the French edition: *Groupes algébriques et corps de classes*. Paris: Hermann, 1975.

Text prepared in camera-ready form using T_EX.
Printed and bound by Braun-Brumfield, Inc., Ann Arbor, MI.
Printed in the United States of America.

9 8 7 6 5 4 3 2 (Corrected second printing, 1997)

ISBN 0-387-96648-X Springer-Verlag New York Berlin Heidelberg
ISBN 3-540-96648-X Springer-Verlag Berlin Heidelberg New York SPIN 10637639

Contents

CHAPTER III
Maps From a Curve to a Commutative Group

CHAPTER IV
Singular Algebraic Curves

CHAPTER V

CHAPTER VI

CHAPTER VII
Group Extension and Cohomology

CHAPTER I

Summary of Main Results

This course presents the work of M. Rosenlicht and S. Lang. We begin by summarizing that of Rosenlicht:

1. Generalized Jacobians

Let X be a projective, irreducible, and non-singular algebraic curve; let $f : X \to G$ be a rational map from X to a commutative algebraic group G. The set S of points of X where f is not regular is a finite set. If D is a divisor prime to S (i.e., of the form $D = \sum n_i P_i$, with $P_i \notin S$), $f(D)$ can be defined to be $\sum n_i f(P_i)$ which is an element of G.

When G is an *Abelian variety*, $S = \emptyset$ and one knows that $f(D) = 0$ if D is the divisor (φ) of a rational function φ on X; in this case, $f(D)$ depends only on the *class* of D for linear equivalence.

In the general case, we are led to modify the notion of class (as in arithmetic, to study ramified extensions) in the following way:

Define a *modulus* with support S to be the data of an integer $n_i > 0$ for each point $P_i \in S$; if $\mathfrak{m}$ is a modulus with support S, and if φ is a rational function, one says that φ is "congruent to 1 mod $\mathfrak{m}$", and one writes $\varphi \equiv 1 \bmod \mathfrak{m}$, if $v_i(1 - \varphi) \geq n_i$ for all i, v_i denoting the valuation attached to the point P_i. Since the n_i are > 0, such a function is regular at the points P_i and takes the value 1 there; its divisor (φ) is thus prime to S.

Theorem 1. *For every rational map $f : X \to G$ regular away from S, there exists a modulus $\mathfrak{m}$ with support S such that $f(D) = 0$ for every divisor $D = (\varphi)$ with $\varphi \equiv 1 \bmod \mathfrak{m}$.*

(For the proof, see chap. III, §2.)

Conversely, given the modulus $\mathfrak{m}$, one can recover, if not the group G, at least a "universal" group for the groups G:

Theorem 2. *For every modulus $\mathfrak{m}$, there exists a commutative algebraic group $J_\mathfrak{m}$ and a rational map $f_\mathfrak{m} : X \to J_\mathfrak{m}$ such that the following property holds:*

For every rational map $f : X \to G$ satisfying the property of theorem 1 with respect to $\mathfrak{m}$, there exists a unique rational (affine) homomorphism $\theta : J_\mathfrak{m} \to G$ such that $f = \theta \circ f_\mathfrak{m}$.

(For the proof, see chap. V, no. 9)

More can be said about the structure of $J_\mathfrak{m}$, exactly as for the usual Jacobian (which we recover if $\mathfrak{m} = 0$). For this, let $C_\mathfrak{m}$ be the group of classes of divisors prime to S modulo those which can be written $D = (\varphi)$ with $\varphi \equiv 1 \bmod \mathfrak{m}$, and let $C_\mathfrak{m}^0$ be the subgroup of $C_\mathfrak{m}$ formed by classes of degree 0. Denoting by C^0 the group of (usual) divisor classes of degree 0, there is a surjective homomorphism $C_\mathfrak{m}^0 \to C^0$. The kernel $L_\mathfrak{m}$ of this homomorphism is formed by the classes in $C_\mathfrak{m}$ of divisors of the form (φ), with φ invertible at each point $P_i \in S$. But, for each $P_i \in S$, the invertible elements modulo those congruent to 1 mod $\mathfrak{m}$ form an algebraic group $R_{\mathfrak{m},i}$ of dimension n_i; let $R_\mathfrak{m}$ be the product of these groups. According to the approximation theorem for valuations, one can find a function corresponding to arbitrary given elements $r_i \in R_{\mathfrak{m},i}$. We conclude that $L_\mathfrak{m}$ is identified with the quotient group $R_\mathfrak{m}/\mathbf{G}_m$, denoting by $\mathbf{G}_m$ the multiplicative group of constants embedded naturally in $R_\mathfrak{m}$. Putting $J = C^0$, we finally have an exact sequence

$$0 \to R_\mathfrak{m}/\mathbf{G}_m \to C_\mathfrak{m}^0 \to J \to 0.$$

Note that J has a natural structure of *algebraic group* since it is the Jacobian of X; the same is true of $R_\mathfrak{m}/\mathbf{G}_m$, as we just saw. This extends to $C_\mathfrak{m}^0$:

Theorem 3. *The map $f_\mathfrak{m} : X \to J_\mathfrak{m}$ defines, by extension to divisor classes, a bijection from $C_\mathfrak{m}^0$ to $J_\mathfrak{m}$. Identifying $C_\mathfrak{m}^0$ and $J_\mathfrak{m}$ by means of this bijection, the group $J_\mathfrak{m}$ becomes an extension (as algebraic group) of the group J by the group $R_\mathfrak{m}/\mathbf{G}_m$.*

(For the proof, see chap. V, §3.)

The groups $J_\mathfrak{m}$ are the *generalized Jacobians* of the curve X.

2. Abelian coverings

Let G be a connected commutative algebraic group, and let $\theta : G' \to G$ be an *isogeny* (the group G' also being assumed connected); recall that this means that θ is a homomorphism (of algebraic groups) which is surjective with a finite kernel. We also suppose that the corresponding field extension is separable, in which case we say that θ is *separable*. If $\mathfrak{g}$ denotes the kernel of θ, the group G is identified with the quotient $G'/\mathfrak{g}$, and G' is an unramified covering of G, with the Abelian group $\mathfrak{g}$ as Galois group.

Now let U be an algebraic variety and let $f : U \to G$ be a regular map. One defines the *pull-back* $U' = f^{-1}(G')$ of G' by f as the subvariety of $U \times G'$ formed by the pairs (x, g') such that $f(x) = \theta(g')$. The projection $U' \to U$ makes U' an (unramified) covering of U, with Galois group $\mathfrak{g}$.

More generally, let $f : X \to G$ be a rational map from an irreducible variety X to the group G, and let $X' \to X$ be a covering of X with Galois group $\mathfrak{g}$. If there exists a non-empty open U of X on which f is regular, and if the covering induced by X' on U is isomorphic to $f^{-1}(G')$, we will again say that X' is the *pull-back* of the isogeny $G' \to G$ by the map f (this amounts to saying that the notion of a pull-back is a birational one).

With this convention, we have:

Theorem 4. *Every Abelian covering of an irreducible algebraic variety is the pull-back of a suitable isogeny.*

We indicate quickly the principle of the proof (for more details, see chap. VI, §2), limiting ourselves to the case of an irreducible covering $X' \to X$. Clearly we can suppose that $\mathfrak{g}$ is a cyclic group of order n, with either n prime to the characteristic, or $n = p^m$.

i) $\mathfrak{g}$ *is cyclic of order* n, *with* $(n, p) = 1$.

Let $\mathbf{G}_m$ be the multiplicative group and let $\theta_n : \mathbf{G}_m \to \mathbf{G}_m$ be the isogeny given by $\lambda \to \lambda^n$. Associating to a generator σ of $\mathfrak{g}$ a primitive n-th root of unity ϵ, we see that the kernel of θ_n is identified with $\mathfrak{g}$. We show that every Abelian covering with Galois group $\mathfrak{g}$ is a pull-back of θ_n:

Let L/K be the field extension corresponding to the given covering $X' \to X$. Since the norm of ϵ in L/K is 1, the classical "theorem 90" of Hilbert shows the existence of $g \in L^*$ such that $g^\sigma = \epsilon.g$, and $L = K(g)$ (the element g is a "Kummer" generator). We have $f = g^n \in K$. The map $g : X' \to \mathbf{G}_m$ commutes with the action of $\mathfrak{g}$ and defines by passage to the quotient the map $f : X \to \mathbf{G}_m$. This shows that $X' = f^{-1}(\mathbf{G}_m)$.

ii) $\mathfrak{g}$ *is cyclic of order* p^m.

First suppose that $m = 1$. Let $\mathbf{G}_a$ be the additive group, and let $\wp : \mathbf{G}_a \to \mathbf{G}_a$ be the isogeny given by $\wp(\lambda) = \lambda^p - \lambda$. The kernel of $\wp$ is the group $\mathbf{Z}/p\mathbf{Z}$ of integers modulo p; choosing a generator σ of $\mathfrak{g}$, it is thus identified with $\mathfrak{g}$. We are going to see that every Abelian covering with Galois group $\mathfrak{g}$ is a pull-back of $\wp$:

Let, as before, L/K be the extension corresponding to the covering. Since the trace of 1 in L/K is 0, the additive analog of "theorem 90" shows the existence of $g \in L$ such that $g^\sigma = g + 1$ (the element g is an "Artin-Schreier" generator) and we have $f = \wp(g) = g^p - g \in K$. As above, this means that the given covering is the pull-back of $\wp$ by g.

When $m > 1$, one replaces $\mathbf{G}_a$ by the group W_m of *Witt vectors* of length m, cf. Witt [99].

Combining theorem 4 with theorems 1 and 2, we get:

Corollary. *Let $X' \to X$ be an Abelian covering of an algebraic curve X. Then there exists a separable isogeny $\theta : G' \to J_\mathfrak{m}$, where $J_\mathfrak{m}$ is a generalized Jacobian of X, such that X' is isomorphic to $f_\mathfrak{m}^{-1}(G')$.*

We also prove the following results (see chap. VI, §2):

a) For fixed X' and $J_\mathfrak{m}$, the isogeny $\theta : G' \to J_\mathfrak{m}$ is *unique*.
b) The modulus $\mathfrak{m}$ can be chosen so that its support S is exactly the set of ramification points of the given covering $X' \to X$.

In particular, unramified coverings correspond to isogenies of the Jacobian.

Using a) and the theorem of "descent of the base field" of Weil [95], we prove (cf. chap. VI, §4):

Theorem 5. *If the Abelian covering $X' \to X$ is defined and Abelian over a finite field k, the isogeny $\theta : G' \to J_\mathfrak{m}$ of the corollary to theorem 4 can be defined over k.*

Thus we get a construction of Abelian extensions of the field $k(X)$ starting from k-isogenies of generalized Jacobians $J_\mathfrak{m}$ corresponding to moduli $\mathfrak{m}$ rational over k. As Lang showed, this construction permits one to easily recover class field theory for the field $k(X)$ (cf. chap VI, §6); in particular, the Artin reciprocity law reduces to a formal calculation in the isogeny θ. The "explicit reciprocity laws" are recovered by means of "local symbols" connected with theorem 1 (see chap. III, §1 as well as chap. VI, no. 30).

3. Other results

a) Class field theory was extended by Lang himself to varieties of any dimension. The maps $f_\mathfrak{m} : X \to J_\mathfrak{m}$ are replaced by "maximal" maps (cf. chap. VI, §3); the most interesting example is that of the canonical map from X to its *Albanese variety*, which furnishes "almost all" of the unramified Abelian extensions of X (cf. chap. VI, no. 20). It should be mentioned that, other than this case and that of curves, one knows very little about maximal maps; one does not know how to extract "generalized Albanese varieties" from them, which would play the role of the $J_\mathfrak{m}$.

b) Other than their arithmetic applications, generalized Jacobians are also interesting as non-trivial *extensions* of an Abelian variety by a linear group.

For example, let $P \in X$ and put $\mathfrak{m} = 2P$. Choosing a local uniformizer t_P at P, we see that the local group $L_\mathfrak{m}$ of no. 1 can be identified with the additive group $\mathbf{G}_a$, and the Jacobian $J_\mathfrak{m}$ is thus an extension of the usual Jacobian J by $\mathbf{G}_a$. By virtue of a result of Rosenlicht (see chap. VII, no. 6), it can be considered as a principal fiber space with base J and group $\mathbf{G}_a$, and thus it defines an element $j_P \in H^1(J, \mathcal{O}_J)$. Let j'_P be the image of j_P by the homomorphism from $H^1(J, \mathcal{O}_J)$ to $H^1(X, \mathcal{O}_X)$ defined by $f_\mathfrak{m}$. Then:

Theorem 6. *Identifying $H^1(X, \mathcal{O}_X)$ with the classes of répartitions on X (cf. chap. II, no. 5), the element $j'_P \in H^1(X, \mathcal{O}_X)$ is identified with the class of the répartition $1/t_P$.*

As we will see, this theorem permits us to determine $H^1(J, \mathcal{O}_J)$, and more generally $H^q(A, \mathcal{O}_A)$ for any Abelain variety A and every integer q (chap. VI, §4).

Bibliographic note

The results summarized above are taken up in the following chapters of this course; at the end of each of these chapters the reader will find a brief bibliographic note. We limit ourselves here to mentioning that the construction and properties of generalized Jacobians are due to Rosenlicht [64], [65] and the arithmetic results of no. 2 are due to Lang [49], [50]; both rely upon the theory of Abelian varieties developed by Weil [89]. The determination of the cohomology of Abelian varieties is essentially due to Rosenlicht [68] and Barsotti [5], [6]; see also [78].

CHAPTER II

Algebraic Curves

In this chapter, as well as the two following ones, we leave aside all questions of rationality. So let us suppose that the base field k is algebraically closed (of any characteristic). For the definitions and elementary results related to algebraic varieties and sheaves, I refer to my memoir on coherent sheaves [73], which will be cited FAC in what follows. In any case, there is no difficulty passing from this language to that of Weil [87], [51], or to that of schemes.

1. Algebraic curves

Let X be an algebraic curve, i.e., an algebraic variety of dimension 1; we will suppose that X is *irreducible*, *non-singular*, and *complete*.

Let $k(X)$ be the field of rational functions on X. It is an extension of finite type of k of transcendence degree 1. Conversely, there is a curve X associated to such an extension F/k, which is unique (up to isomorphism).

First we show the *existence* of X. Let $x_1, \ldots, x_r$ be generators of the extension F/k and let $A = k[x_1, \ldots, x_r]$ be the subalgebra of F generated by the x_i; it is an affine algebra, corresponding to a closed subvariety Y of the affine space k^r. Its closure $\overline{Y}$ in the projective space $\mathbf{P}_r(k)$ is a complete irreducible curve whose field of rational functions is F. To find the curve X, it then suffices to take the *normalization* of $\overline{Y}$; indeed, one knows that a normal curve is non-singular. Furthermore, the method of projective normalization ([71], pp. 25-26 or [51], pp. 133-146, for example) shows that X *can be embedded in a projective space*.

The *uniqueness* of X follows from the explicit determination of its Zariski topology and its local rings, cf. no. 2; moreover, one knows that the knowledge of the local rings of an irreducible variety X of any dimension determines the Zariski topology of X, cf. [17], exposés 1 and 2.

(The uniqueness of X can also be deduced from the following fact: every rational map from a non-singular curve to a complete variety is everywhere regular.)

The study of X is thus *equivalent* to the study of the extension F/k, contrary to what could happen for a variety of dimension ≥ 2. There is thus no reason to insist on the difference between "geometric" methods and "algebraic" methods.

2. Local rings

Let P be a point of the curve X. One knows how to define the *local ring* $\mathcal{O}_P$ of X at P: supposing that X is embedded in a projective space $\mathbf{P}_r(k)$, it is the set of functions induced by rational functions of the type R/S, where R and S are homogeneous polynomials of the same degree and where $S(P) \neq 0$. It is a subring of $k(X)$; by virtue of the general properties of algebraic varieties, it is a *Noetherian local ring* whose maximal ideal $\mathfrak{m}_P$ is formed by the functions f vanishing at P and we have $\mathcal{O}_P/\mathfrak{m}_P = k$. The elements of $\mathcal{O}_P$ will be called *regular at P*.

Now let us use the hypotheses made on X. Since X is a curve, $\mathcal{O}_P$ is a local ring of *dimension 1*, in the sense of dimension theory for local rings: its only prime ideals are (0) and $\mathfrak{m}_P$. Since P is a simple point of X, it is also a *regular* local ring: its maximal ideal can be generated by a single element; such an element t will be called a *local uniformizer* at P. By virtue of a well-known (and elementary) theorem, these properties imply that $\mathcal{O}_P$ is a *discrete valuation ring*; the corresponding valuation will be written v_P. If f is a non-zero element of $k(X)$, the relation $v_P(f) = n$, $n \in \mathbf{Z}$ thus means that f can be written in the form $f = t^n u$ where t is a local uniformizer at P and u is an invertible element of $\mathcal{O}_P$. Furthermore, the rings $\mathcal{O}_P$ are the only valuation rings of $k(X)$ containing k; indeed, if U is such a ring, U dominates one of the $\mathcal{O}_P$ (since X is assumed complete—this is one of the definitions of a complete variety, cf. [11]), thus coincides with $\mathcal{O}_P$ since the latter is a valuation ring.

As with any algebraic variety, the $\mathcal{O}_P$ form a *sheaf of rings* on X when X is given the *Zariski topology* (FAC, chap. II); recall that the closed subsets in this topology are the finite subsets and X itself. The sheaf $\mathcal{O}_P$ will be denoted $\mathcal{O}_X$ or simply $\mathcal{O}$ when no confusion can result; it is a subsheaf of the constant sheaf $k(X)$.

3. Divisors, linear equivalence, linear series

An element of the free Abelian group on the points $P \in X$ is called a *divisor*. A divisor is thus written

$$D = \sum_{P \in X} n_P P \qquad \text{with } n_P \in \mathbf{Z},$$

and $n_P = 0$ for almost all P (all but finitely many). The coefficient of P in D will be written $v_P(D)$.

The *degree* of D is defined by

$$\deg(D) = \sum n_P = \sum v_P(D)$$

A divisor D is called *effective* (or *positive*) if all the $v_P(D)$ are ≥ 0; thus there is an order structure on the group $D(X)$ of all divisors on X.

If f is a non-zero element of $k(X)$, one defines the *divisor of f*, written (f), by the formula

$$(f) = \sum_{P \in X} v_P(f) P.$$

By virtue of the evident identity $(fg) = (f) + (g)$, these divisors form a subgroup $P(X)$ of the group $D(X)$ as f runs through $k(X)^*$. The quotient group $C(X) = D(X)/P(X)$ is called the group of *divisor classes* (for linear equivalence) and two divisors in the same class are said to be linearly equivalent.

Proposition 1. *If $D \in P(X)$, then $\deg(D) = 0$.*

PROOF. This result is an immediate consequence of the Riemann-Roch theorem in its first form (no. 4), which we will prove without using it. But we can also give a direct proof: if $D = (f)$, with $f \in k(X)^*$, we can suppose that f is non-constant (otherwise $D = 0$). The function f is then a map from X to the projective line $\mathbf{P}_1(k)$, and (f) is nothing other than $f^{-1}(0) - f^{-1}(\infty)$, 0 and ∞ being identified with two points of $\mathbf{P}_1(k)$, and the operation f^{-1} being taken in the sense of intersection theory. But one knows (thanks to this same theory) that, for every point $a \in \mathbf{P}_1(k)$, the degree of $f^{-1}(a)$ is equal to the degree of the projection f, i.e., to $[k(X) : k(f)]$. Whence the proposition, with added precision (which shows, for example, that neither $f^{-1}(0)$ nor $f^{-1}(\infty)$ are reduced to 0 for a non-constant function f—in other words, the inequality $(f) \geq 0$ implies that f is constant).   □

It follows from prop. 1 that one can speak of the *degree* of a divisor class, and in particular of the group $C^0(X)$ of divisor classes of degree 0. We get

$$C(X)/C^0(X) = \mathbf{Z}.$$

Combining linear equivalence with the order relation on divisors, we arrive at the notion of a *linear series*:

Let D be any divisor, and consider the divisors D' which are effective and linearly equivalent to D. Such a divisor can be written $D' = D + (f)$, with $f \in k(X)^*$, and we must have $D + (f) \geq 0$, i.e., $(f) \geq -D$. The functions f satisfying this condition, together with 0, form a vector space which will be written $L(D)$. We will see later (prop. 2) that $L(D)$ is finite dimensional. Every element $f \neq 0$ of $L(D)$ defines a divisor $D' = D + (f)$ of the type considered, and two functions f and g define the same divisor if and only if $f = \lambda g$ with $\lambda \in k^*$; thus, the set $|D|$ of effective divisors linearly equivalent to D is in bijective correspondence with the *projective space* $\mathbf{P}(L(D))$ associated to the vector space $L(D)$. The structure of projective space thus defined on $|D|$ does not change when D is replaced by a linearly equivalent divisor. A non-empty set F of effective divisors on X is called a *linear series* if there exists a divisor D such that F is a projective (linear) subvariety of $|D|$; if $F = |D|$, one says that the linear series F is *complete*. A linear series F, contained in $|D|$, corresponds to a vector subspace V of $L(D)$; the dimension of V is equal to the (projective) dimension of F plus 1. In particular, if $l(D)$ denotes the dimension of $L(D)$, then

$$l(D) = \dim |D| + 1.$$

Remark. Linear series are closely related to maps of X to a projective space. We indicate rapidly how:

Let $\varphi : X \to \mathbf{P}_r(k)$ be a regular map from X to a projective space. We suppose that $\varphi(X)$ generates (projectively) $\mathbf{P}_r(k)$. With this hypothesis, if H denotes a hyperplane of $\mathbf{P}_r(k)$, the divisor $\varphi^{-1}(H)$ is well-defined. One immediately checks that, as H varies, the $\varphi^{-1}(H)$ form a linear series F of dimension r, "without fixed points" (i.e., for every $P \in X$ there exists $D \in F$ such that $v_P(D) = 0$); conversely, every linear series without fixed points arises uniquely (up to an automorphism of $\mathbf{P}_r(k)$) this way. Furthermore, for every linear series F there exists an effective divisor A and a linear series F' without fixed points such that F is the set of divisors of the form $A + D'$, where D' runs through F'; the divisor A is called the *fixed part* of F.

(This discussion extends, with evident modifications, to the case where X is a normal variety of any dimension. However, one must distinguish between the *fixed components* of a linear series F (these are the subvarieties W of X, of codimension 1, such that $D \geq W$ for all $D \in F$) and the *base points* of F (these are the points of intersection of the supports of the divisors $D \in F$). The rational map from X to the projective space associated to F does not change when the fixed components are removed from F; this map is regular away from the base points of F. For more details, see for example Lang [51], chap. VI.)

4. The Riemann-Roch theorem (first form)

Let D be a divisor on X. In the preceding no. we defined the vector space
$L(D)$: it is the set of rational functions f which satisfy $(f) \geq -D$, that is
to say

$$v_P(f) \geq -v_P(D) \qquad \text{for all } P \in X.$$

Now if P is a point of X, write $\mathcal{L}(D)_P$ for the set of functions which satisfy
this inequality at P. The $\mathcal{L}(D)_P$ form a *subsheaf* $\mathcal{L}(D)$ of the constant
sheaf $k(X)$. The group $H^0(X, \mathcal{L}(D))$ is just $L(D)$.

Proposition 2. *The vector spaces $H^0(X, \mathcal{L}(D))$ and $H^1(X, \mathcal{L}(D))$ are
finite dimensional over k. For $q \geq 2$, $H^q(X, \mathcal{L}(D)) = 0$.*

PROOF. According to FAC, no. 53, $H^q(X, \mathcal{F}) = 0$ for $q \geq 2$ and any
sheaf $\mathcal{F}$, whence the second part of the proposition. To prove the first
part it suffices, according to FAC, no. 66, to prove that $\mathcal{L}(D)$ is a coherent
algebraic sheaf. But, if P is a point of X and φ a function such that
$v_P(\varphi) = v_P(D)$, one immediately checks that multiplication by φ is an
isomorphism from $\mathcal{L}(D)$ to the sheaf $\mathcal{O}$ in a neighborhood of P; *a fortiori*,
$\mathcal{L}(D)$ is coherent. $\square$

Remarks. 1. If $D' = D + (\varphi)$, the sheaf $\mathcal{L}(D)$ is isomorphic to the sheaf
$\mathcal{L}(D')$, the isomorphism being defined by multiplication by φ.

2. It would be easy to prove prop. 2 without using the results of FAC by
using the direct definitions of $H^0(X, \mathcal{L}(D))$ and $H^1(X, \mathcal{L}(D))$; for this see
the works cited at the end of the chapter.

Before stating the Riemann-Roch theorem, we introduce the following
notations:

$$I(D) = H^1(X, \mathcal{L}(D)), \quad i(D) = \dim I(D), \quad g = i(0) = \dim H^1(X, \mathcal{O}).$$

The integer g is called the *genus* of the curve X; we will see later that this
definition is equivalent to the usual one.

Theorem 1 (Riemann-Roch theorem—first form). *For every divisor D,*
$$l(D) - i(D) = \deg(D) + 1 - g.$$

PROOF. First observe that this formula is true for $D = 0$. Indeed, $l(0) =
1$ (because, as we saw, the constants are the only functions f satisfying
$(f) \geq 0$), $i(0) = g$ by definition, and $\deg(0) = 0$.

It will thus suffice to show that, if the formula is true for a divisor D,
it is true for $D + P$, and conversely (P being any point of X); indeed, it
is clear that one can pass from the divisor 0 to any divisor by succesively
adding or subtracting a point.

Denote the left hand side of the formula by $\chi(D)$ and the right hand side by $\chi'(D)$; evidently $\chi'(D + P) = \chi'(D) + 1$ and thus we must show that the same formula holds for $\chi(D)$. But, the sheaf $\mathcal{L}(D)$ is a subsheaf of $\mathcal{L}(D + P)$, which permits us to write an exact sequence

$$0 \to \mathcal{L}(D) \to \mathcal{L}(D + P) \to \mathcal{Q} \to 0.$$

The quotient sheaf $\mathcal{Q}$ is zero away from P, and $\mathcal{Q}_P$ is a vector space of dimension 1. Thus $H^1(X, \mathcal{Q}) = 0$ and $H^0(X, \mathcal{Q}) = \mathcal{Q}_P$ is a vector space of dimension 1. We write the cohomology exact sequence

$$0 \to L(D) \to L(D + P) \to H^0(X, \mathcal{Q}) \to I(D) \to I(D + P) \to 0.$$

Taking the alternating sum of the dimensions of these vector spaces we find

$$l(D) - l(D + P) + 1 - i(D) + i(D + P) = 0,$$

that is to say

$$\chi(D + P) = \chi(D) + 1,$$

as was to be shown.   □

Remarks. 1. Theorem 1 is not enough to "compute" $l(D)$: one must also have information about $i(D)$. This information will be furnished by the *duality theorem* (no. 8) and we will then obtain the definitive form of the Riemann-Roch theorem.

2. The method of proof above, consisting of checking the theorem for *one* divisor, then passing from one divisor to another by means of the sheaf $\mathcal{Q}$ supported on a subvariety, also applies to varieties of higher dimension. For example, it is not difficult to prove in this way the Riemann-Roch theorem for a non-singular surface in the form

$$\chi(D) = \frac{1}{2}D(D - K) + 1 + p_a,$$

K denoting the canonical divisor and p_a the arithmetic genus of the surface under consideration. (See chap. IV, no. 8.)

5. Classes of répartitions

Before passing to differentials and the duality theorem, we are going to show how the vector space $I(D)$ can be interpreted in Weil's language of *répartitions* (or "adèles").

A répartition r is a family $\{r_P\}_{P \in X}$ of elements of $k(X)$ such that $r_P \in \mathcal{O}_P$ for almost all $P \in X$. The répartitions form an algebra R over the field k. If D is a divisor, we write $R(D)$ for the vector subspace of R formed by the $r = \{r_P\}$ such that $v_P(r_P) \geq -v_P(D)$; as D runs through the ordered set of divisors of X, the $R(D)$ form an increasing filtered family of subspaces of R whose union is R itself.

On the other hand, if to every $f \in k(X)$ we associate the répartition $\{r_P\}$ such that $r_P = f$ for every $P \in X$, we get an injection of $k(X)$ into R which permits us to identify $k(X)$ with a subring of R. With these notations, we have:

Proposition 3. *If D is a divisor on X, then the vector space $I(D) = H^1(X, \mathcal{L}(D))$ is canonically isomorphic to $R/(R(D) + k(X))$.*

PROOF. The sheaf $\mathcal{L}(D)$ is a subsheaf of the constant sheaf $k(X)$. Thus there is an exact sequence

$$0 \to \mathcal{L}(D) \to k(X) \to k(X)/\mathcal{L}(D) \to 0.$$

As the curve X is irreducible and the sheaf $k(X)$ is constant,

$$H^1(X, k(X)) = 0$$

(since the nerve of every open cover of X is a simplex); on the other hand, since X is connected, $H^0(X, k(X)) = k(X)$. Thus the cohomology exact sequence associated to the exact sequence of sheaves above can be written

$$k(X) \to H^0(X, k(X)/\mathcal{L}(D)) \to H^1(X, \mathcal{L}(D)) \to 0.$$

The sheaf $\mathcal{A} = k(X)/\mathcal{L}(D)$ is a "sky-scraper sheaf": if s is a section of $\mathcal{A}$ over a neighborhood U of a point P, there exists a neighborhood $U' \subset U$ of the point P such that $s = 0$ on $U' - P$. It follows that $H^0(X, \mathcal{A})$ is identified with the direct sum of the $\mathcal{A}_P$ for $P \in X$; but this direct sum is visibly isomorphic to $R/R(D)$. The exact sequence written above then shows that $H^1(X, \mathcal{L}(D))$ is identified with $R/(R(X) + k(X))$, as was to be shown. $\square$

In all that follows, we identify $I(D)$ and $R/(R(X) + k(X))$.

6. Dual of the space of classes of répartitions

The notations being the same as those in the preceding no., let $J(D)$ be the dual of the vector space $I(D) = R/(R(D) + k(X))$; an element of $J(D)$ is thus identified with a linear form on R, vanishing on $k(X)$ and on $R(D)$. If $D' \geq D$, then $R(D') \supset R(D)$, which shows that $J(D) \supset J(D')$. The union of the $J(D)$, for D running through the set of divisors of X, will be denoted J; observe that the family of the $J(D)$ is a decreasing filtered family.

(One can also interpret J as the *topological dual* of $R/k(X)$ where $R/k(X)$ is given the topology defined by the vector subspaces which are the images of the $R(D)$.)

Let $f \in k(X)$ and let $\alpha \in J$. The map $r \to \langle \alpha, fr \rangle$ is a linear form on R, vanishing on $k(X)$; we denote it by $f\alpha$. We have $f\alpha \in J$; indeed, if $\alpha \in J(D)$ and $f \in L(\Delta)$, we immediately see that the linear form $f\alpha$ vanishes on $R(D-\Delta)$, thus belongs to $J(D-\Delta)$. The operation $(f, \alpha) \to f\alpha$ endows J with the structure of *vector space* over $k(X)$.

Proposition 4. *The dimension of the vector space J over the field $k(X)$ is ≤ 1.*

PROOF. We argue by contradiction; let α and α' be two elements of J which are linearly independent over $k(X)$. Since J is the union of the filtered set of the $J(D)$, one can find a D such that $\alpha \in J(D)$ and $\alpha' \in J(D)$; put $d = \deg(D)$.

For every integer $n \geq 0$, let Δ_n be a divisor of degree n (for example $\Delta_n = nP$, where P is a fixed point of X). If $f \in L(\Delta_n)$, then $f\alpha \in J(D - \Delta_n)$ in light of what was said above, and similarly for $g\alpha'$ if $g \in L(\Delta_n)$. Furthermore, since α and α' are linearly independent over $k(X)$, the relation $f\alpha + g\alpha' = 0$ implies $f = g = 0$; it follows that the map

$$(f, g) \to f\alpha + g\alpha'$$

is an injection from the direct sum $L(\Delta_n) + L(\Delta_n)$ to $J(D - \Delta_n)$, and in particular we have the inequality

$$\dim J(D - \Delta_n) \geq 2 \dim L(\Delta_n) \qquad \text{for all } n. \qquad (*)$$

We are going to show that the inequality $(*)$ leads to a contradiction when $n \to +\infty$. The left hand side is equal to

$$\dim I(D - \Delta_n) = i(D - \Delta_n).$$

According to thm. 1,

$$i(D - \Delta_n) = -\deg(D - \Delta_n) + g - 1 + l(D - \Delta_n)$$
$$= n + (g - 1 - d) + l(D - \Delta_n).$$

But when $n > d$, $\deg(D - \Delta_n) < 0$, which evidently implies

$$l(D - \Delta_n) = 0$$

(indeed, otherwise there would exist an effective divisor linearly equivalent to $D - \Delta_n$, which is impossible in view of prop. 1). Thus for large n the left hand side of $(*)$ is equal to $n + A_0$, A_0 being a constant.

As for the right hand side, it is equal to $2l(\Delta_n)$. Thm. 1 shows that

$$l(\Delta_n) \geq \deg(\Delta_n) + 1 - g = n + 1 - g.$$

Thus the right hand side of $(*)$ is $\geq 2n + A_1$, A_1 denoting a constant, and we get a contradiction for n sufficiently large, as was to be shown. $\square$

Remarks. 1. It would be easy to show that the dimension of J is exactly 1: it would suffice to exhibit a non-zero element of J. In fact, we will prove later a more precise result, namely that J is isomorphic to the space of *differentials* on X.

2. The definitions and results of this no. can be easily transposed to the case of a normal projective variety of any dimension r: if D is a divisor on X, one again defines $J(D)$ as the dual of $H^r(X, \mathcal{L}(D))$. From the fact that all the H^{r+1} are zero, the exact sequence of cohomology shows that the functor H^r is right exact, and, if $D' \geq D$, one again has an injection from $J(D')$ to $J(D)$. The inductive limit J of the $J(D)$ is a vector space over $k(X)$ of dimension 1: this is seen by an argument analogous to that of prop. 4 (one must take, in place of Δ_n, a multiple of the hyperplane section of X); the only results of sheaf theory that we have used are the very elementary ones of FAC, no. 66. (For more details, see the report of Zariski [103], p. 139.)

7. Differentials, residues

Recall briefly the general notion of a *differential* on an algebraic variety X:

First of all, if F is a commutative algebra over a field k, we have the *module of k-differentials of F*, written $D_k(F)$; it is an F-module, endowed with a k-linear map

$$d : F \to D_k(F),$$

satisfying the usual condition $d(xy) = x.dy + y.dx$. The dx for $x \in F$ generate $D_k(F)$ and $D_k(F)$ is the "universal" module with these properties. For more details, see [11], exposé 13 (Cartier).

These remarks apply in particular to the local rings $\mathcal{O}_P$ and to the field of rational functions $F = k(X)$ of an algebraic variety X (of any dimension r). Reducing to the affine case, one immediately checks that the $\Omega_P = D_k(\mathcal{O}_P)$ form a coherent algebraic sheaf on X; furthermore

$$D_k(F) = D_k(\mathcal{O}_P) \otimes_{\mathcal{O}_P} F.$$

If P is a simple point of X and if $t_1, \ldots, t_r$ form a regular system of parameters at P, the dt_i form a basis of $D_k(\mathcal{O}_P)$; this can be seen, for example, by applying thm. 5 of exposé 17 of the Seminar cited above. Thus the sheaf of Ω_P is locally free over the open set of simple points of X (it thus corresponds to a vector bundle which is nothing other than the dual of the tangent space).

Now if we come back to the case of a *curve* satisfying the conditions of no. 1, we see that, in this case, $D_k(F)$ is a vector space of dimension 1 over $F = k(X)$ and that the sheaf Ω of the Ω_P is a subsheaf of the constant sheaf $D_k(F)$. If t is a local uniformizer at P, the differential dt of t is a basis of the $\mathcal{O}_P$-module Ω_P and it is also a basis of the F-vector space $D_k(F)$. Thus if $\omega \in D_k(F)$, we can write $\omega = f\,dt$, with $f \in F$. Then supposing $\omega \neq 0$, we put

$$v_P(\omega) = v_P(f).$$

One sees immediately that this definition is indeed *invariant*, i.e., independent of the choice of dt; moreover, it would apply to any rational section of a line bundle (i.e., of a vector bundle of fiber dimension 1).

From the expression $\omega = f\,dt$, we can also deduce another local invariant of ω, its *residue*: if $\widehat{F}_P$ denotes the completion of the field F for the valuation v_P, one knows that $\widehat{F}_P$ is isomorphic to the field $k((T))$ of formal series over k, the isomorphism being determined by the condition that t maps to T. Identifying f with its image in $\widehat{F}_P$, we can thus write

$$f = \sum_{n >> -\infty} a_n T^n, \qquad a_n \in k.$$

the symbol $n >> -\infty$ meaning that n only takes a finite number of values < 0.

In particular, the coefficient a_{-1} of T^{-1} in f is well defined, and it is this coefficient which will be called the *residue* of $\omega = f\,dt$ at P, written $\mathrm{Res}_P(\omega)$. This definition is justified by the following proposition:

Proposition 5 (Invariance of the residue). *The preceding definition is independent of the choice of the local uniformizer t.*

The proof will be given later (no. 11) at the same time as the list of the properties of the operation $\omega \to \mathrm{Res}_P(\omega)$. Just note for the moment that $\mathrm{Res}_P(\omega) = 0$ if $v_P(\omega) \geq 0$, i.e., if ω does not have a pole at P. As every differential has only a finite number of poles (since it is a rational section of a vector bundle), we conclude that $\mathrm{Res}_P(\omega) = 0$ for almost all P and the sum $\sum_{P \in X} \mathrm{Res}_P(\omega)$ makes sense. On this subject we have the following fundamental result:

Proposition 6 (Residue formula). *For every differential $\omega \in D_k(F)$, $\sum_{P \in X} \mathrm{Res}_P(\omega) = 0$.*

The proof will be given later (nos. 12 and 13). This proof, as well as that of prop. 5, is very simple when the characteristic is zero, but is much less so in characteristic $p > 0$. However, in the latter case one can give proofs of a different character, using the operation defined by Cartier, cf. [12].

As for the case of characteristic 0, one can also, of course, treat it by "transcendental" techniques. Indeed, according to the Lefschetz principle, we can suppose that $k = \mathbf{C}$, the field of complex numbers; the curve X can then naturally be given a structure of a compact complex analytic variety of dimension 1. One immediately checks that $\mathrm{Res}_P(\omega) = \frac{1}{2\pi i} \oint_P \omega$, which proves prop. 5; as for prop. 6, it follows from Stokes' formula.

8. Duality theorem

Let ω be a non-zero differential on the curve X. We define its *divisor* (ω) by the same formula as in the case of functions:

$$(\omega) = \sum_{P \in X} v_P(\omega)P, \qquad v_P(\omega) \text{ defined as in the preceding no.}$$

If D is a divisor, we write $\Omega(D)$ for the vector space formed by 0 and the differentials $\omega \neq 0$ such that $(\omega) \geq D$; it is a subspace of the space $D_k(F)$ of all differentials on X.

Given these definitions, we are going to define a scalar product $\langle \omega, r \rangle$ between differentials $\omega \in D_k(F)$ and répartitions $r \in R$ by means of the following formula:

$$\langle \omega, r \rangle = \sum_{P \in X} \mathrm{Res}_P(r_P \omega).$$

This definition is legitimate since $r_P \omega \in \Omega_P$ for almost all P. The scalar product thus defined has the following properties:

a) $\langle \omega, r \rangle = 0$ if $r \in F = k(X)$, because of the residue formula (prop. 6).
b) $\langle \omega, r \rangle = 0$ if $r \in R(D)$ and $\omega \in \Omega(D)$ for then $r_P \omega \in \underline{\Omega}_P$ for every $P \in X$.
c) If $f \in F$, then $\langle f\omega, r \rangle = \langle \omega, fr \rangle$.

For every differential ω, let $\theta(\omega)$ be the linear form on R which sends r to $\langle \omega, r \rangle$. Properties a) and b) mean that, if $\omega \in \Omega(D)$, then $\theta(\omega) \in J(D)$ since $J(D)$ is by definition the dual of $R/(R(D) + k(X))$.

Theorem 2 (Duality theorem). *For every divisor D, the map θ is an isomorphism from $\Omega(D)$ to $J(D)$.*

(In other words, the scalar product $\langle \omega, r \rangle$ puts the vector spaces $\Omega(D)$ and $I(D) = R/(R(D) + k(X))$ in duality.)

First we prove a lemma:

Lemma 1. *If ω is a differential such that $\theta(\omega) \in J(D)$, then $\omega \in \Omega(D)$.*

PROOF. Indeed, otherwise there would be a point $P \in X$ such that $v_P(\omega) < v_P(D)$. Put $n = v_P(\omega) + 1$, and let r be the répartition whose components are

$$\begin{cases} r_Q = 0 \text{ if } Q \neq P, \\ r_P = 1/t^n, \ t \text{ being a local uniformizer at } P. \end{cases}$$

We have $v_P(r_P\omega) = -1$, whence $\mathrm{Res}_P(r_P\omega) \neq 0$ and $\langle \omega, r \rangle \neq 0$; but since $n \leq v_P(D)$, $r \in R(D)$ and we arrive at a contradiction since $\theta(\omega)$ is assumed to vanish on $R(D)$. $\qquad\square$

We can now prove thm. 2. First of all, θ is *injective*. Indeed, if $\theta(\omega) = 0$, the preceding lemma shows that $\omega \in \Omega(\Delta)$ for every divisor Δ, whence evidently $\omega = 0$. Next, θ is *surjective*. Indeed, according to c), θ is an *F-linear* map from $D_k(F)$ to J; as $D_k(F)$ has dimension 1, and J has dimension ≤ 1 (prop. 4), θ maps $D_k(F)$ onto J. Thus if α is any element of $J(D)$, there exists $\omega \in D_k(F)$ such that $\theta(\omega) = \alpha$, and the lemma above shows that $\omega \in \Omega(D)$. $\square$

Corollary. *We have $i(D) = \dim \Omega(D)$. In particular, the genus $g = i(0)$ is equal to the dimension of the vector space of differential forms such that $(\omega) \geq 0$ (forms "of the first kind").*

Thus we recover the usual definition of the genus.

9. The Riemann-Roch theorem (definitive form)

Let ω and ω' be two differentials $\neq 0$. Since $D_k(F)$ has dimension 1 over F, we have $\omega' = f\omega$ with $f \in F^*$, whence $(\omega') = (f) + (\omega)$. Thus, all divisors of differential forms are linearly equivalent and form a single class for linear equivalence, called the *canonical class* and written K. By abuse of language, one often writes K for a divisor belonging to this class.

Now let D be any divisor; we seek to determine $\Omega(D)$. If $K = (\omega_0)$ is a canonical divisor, every differential ω can be written $\omega = f\omega_0$ and $(\omega) \geq D$ if and only if $(f) + (\omega_0) \geq D$, i.e., if $f \in L(K - D)$. We conclude that

$$i(D) = \dim \Omega(D) = l(K - D),$$

and, combining this result with thm. 1, we finally get:

Theorem 3 (Riemann-Roch theorem—definitive form). *For every divisor D, $l(D) - l(K - D) = \deg(D) + 1 - g$.*

We put $D = K$ in this formula. Then $l(K) = i(0) = g$ and $l(0) = 1$, whence $g - 1 = \deg(K) + 1 - g$, and we get

$$\deg(K) = 2g - 2.$$

Corollary.
a) *If $\deg(D) \geq 2g - 1$, then the complete linear series $|D|$ has dimension $\deg(D) - g$.*
b) *If $\deg(D) \geq 2g$, $|D|$ has no fixed points.*
c) *If $\deg(D) \geq 2g + 1$, $|D|$ is ample—that is to say it defines a biregular embedding of X in a projective space.*

PROOF. If $\deg(D) \geq 2g - 1$, then $\deg(K - D) \leq -1$, whence $l(K - D) = 0$ and $l(D) = \deg(D) + 1 - g$, which proves a).

Now suppose that $\deg(D) \geq 2g$ and that $|D|$ has a fixed point P. Then there exists a linear series F such that the divisors of $|D|$ are of the form $P + H$ where H runs through F. Thus $\dim F = \dim |D|$, which contradicts a) since $\deg(F) = \deg(D) - 1$.

Finally, suppose that $\deg(D) \geq 2g + 1$, and let $P \in X$. According to b), the linear series $|D - P|$ has no fixed points. Thus there exists $\Delta \in D$ such that $v_P(\Delta) = 1$. If $\varphi : X \to \mathbf{P}_r(k)$ is the map associated to $|D|$ (cf. no. 3), this means that there exists a hyperplane H of $\mathbf{P}_r(k)$ such that $\varphi^{-1}(H)$ contains P with coefficient 1. It follows first of all that the map $\varphi : X \to \varphi(X)$ has degree 1, then that $\varphi(P)$ is a simple point of $\varphi(X)$. The map φ is thus an isomorphism, as was to be shown. $\square$

For other applications of the Riemann-Roch theorem (to "Weierstrass points" for example), see the treatise of Severi [79].

10. Remarks on the duality theorem

Since $i(K) = l(0) = 1$, the vector space $H^1(X, \mathcal{L}(K))$ is one dimensional; the same is thus true of $H^1(X, \Omega)$ since the sheaf Ω is isomorphic to the sheaf $\mathcal{L}(K)$. In fact, making explicit this last isomorphism as well as the duality between $H^1(X, \mathcal{L}(K))$ and $\Omega(K) = L(0)$, one sees that $H^1(X, \Omega)$ has a *canonical* basis, in other words it is *canonically isomorphic to k*.

The scalar product $\langle \omega, r \rangle$ between the elements of $\Omega(D) = H^0(X, \Omega(D))$ and $I(D) = H^1(X, \mathcal{L}(D))$ can then be interpreted as a *cup-product* with values in $H^1(X, \Omega)$ and the duality theorem says that this product puts the two spaces in duality. In this form, the theorem can be extended to an arbitrary coherent algebraic sheaf $\mathcal{F}$: putting $\widetilde{\mathcal{F}} = \mathrm{Hom}_\mathcal{O}(\mathcal{F}, \Omega)$, the cup product maps $H^1(X, \mathcal{F}) \times H^0(X, \widetilde{\mathcal{F}})$ to $H^1(X, \Omega)$ and puts the two spaces in duality.

We also mention that thm. 2, as well as its proof, extends without great modification to normal varieites of any dimension r. The sheaf Ω should then be replaced by the sheaf Ω^r of differential forms of degree r without poles; one proves by induction on r that $H^r(X, \Omega^r)$ is canonically isomorphic to k. Given this, the cup-product defines a scalar product on $H^1(X, \mathcal{L}(D)) \times H^0(X, \Omega^r(D))$, whence a linear map θ from $H^0(X, \Omega^r(D))$ to the dual $J(D)$ of $H^r(X, \mathcal{L}(D))$. The argument of thm. 2 then shows that θ is an isomorphism. For more details, see the report of Zariski already cited.

11. Proof of the invariance of the residue

The rest of this chapter contains the proof of propositions 5 and 6, stated in no. 7. We begin with proposition 5:

This is a *local* question, bearing on differentials of the field $\widehat{F}_P$; this field will be denoted by K in the rest of this no. The choice of a local uniformizer t identifies K with $k((t))$. We will denote by v the valuation of K, by $\mathcal{O}$ its valuation ring (the set of $f \in K$ such that $v(f) \geq 0$), and by $\mathfrak{m}$ its maximal ideal (the set of $f \in K$ such that $v(f) > 0$); evidently $\mathcal{O} = \widehat{\mathcal{O}}_P$ and $\mathfrak{m} = \widehat{\mathfrak{m}}_P$.

The module $D_k(K)$ of differentials of K is defined by the procedure indicated in no. 7. Because this procedure does not take into account the valuation of K, we get (in characteristic 0) a module that is "too large": it is an infinite-dimensional vector space over K. It is convenient to pass to the associated separated module (for the $\mathfrak{m}$-topology), by putting

$$D'_k(K) = D_k(K)/Q \qquad \text{with} \qquad Q = \bigcap_{n \geq 0} \mathfrak{m}^n d(\mathcal{O}).$$

This module no longer has pathological properties:

Lemma 2. *Let t be a local uniformizer and for every element $f = \sum_{n \gg -\infty} a_n t^n$ of K put $f'_t = \sum_{n \gg -\infty} n a_n t^{n-1}$. Then $df = f'_t \, dt$ in $D'_k(K)$ and dt forms a basis of $D'_k(K)$ over K.*

PROOF. To show that $df = f'_t \, dt$ in $D'_k(K)$ we must prove that, for every integer $N \geq 0$, $df - f'_t \, dt \in \mathfrak{m}^N d(\mathcal{O})$ in $D_k(K)$. This presents no problems; we write

$$f = f_0 + t^{N+1} f_1 \qquad \text{with} \qquad f_0 = \sum_{n < N} a_n t^n, \quad f_1 \in \mathcal{O},$$

$$f'_t = (f_0)'_t + t^N g, \qquad \text{with} \qquad g \in \mathcal{O},$$

and we find

$$df - f'_t \, dt = (N+1) t^N f_1 \, dt + t^{N+1} df_1 - t^N g \, dt,$$

and as these three terms belong to $\mathfrak{m}^N d(\mathcal{O})$ this proves the first assertion of the lemma.

Thus dt generates the K-vector space $D'_k(K)$; to show that it is a basis it suffices to prove that $D'_k(K) \neq 0$, that is to say that there exists a derivation of K, not identically zero, and whose extension to $D_k(K)$ vanishes on Q. The derivation $D : K \to K$ defined by $Df = f'_t$ has these properties; indeed, it is not identically zero, and it maps $\mathfrak{m}^{N+1} d(\mathcal{O})$ to $\mathfrak{m}^N$, thus it maps Q to $\bigcap \mathfrak{m}^N$, which is zero. $\qquad \square$

From now on, by a *differential* of K we mean an element of $D'_k(K)$; if ω is such a differential and if t is a local uniformizer, then $\omega = f \, dt$, with

$f \in K$. If $f = \sum a_n t^n$, the coefficient a_{-1} of dt/t in ω will be called the *residue* of ω (with respect to t) and written $\mathrm{Res}_t(\omega)$. Proposition 5 can then be reformulated in the following manner:

Proposition 5′. *If t and u are two local uniformizers of K, then $\mathrm{Res}_t(\omega) = \mathrm{Res}_u(\omega)$ for every differential $\omega \in D'_k(K)$.*

We first note some properties of the operation $\mathrm{Res}_t(\omega)$:

i) $\mathrm{Res}_t(\omega)$ is k-linear in ω.
ii) $\mathrm{Res}_t(\omega) = 0$ if $v(\omega) \geq 0$ (i.e., if $\omega \in \mathcal{O}\, dt$).
iii) $\mathrm{Res}_t(dg) = 0$ for every $g \in K$.
iv) $\mathrm{Res}_t(dg/g) = v(g)$ for every $g \in K^*$.

Properties i,) ii), and iii) are evident. For iv), put $g = t^n w$ with $n = v(g)$, so $v(w) = 0$. Then we find

$$dg/g = n\, dt/t + dw/w,$$

whence $\mathrm{Res}_t(dg/g) = n + \mathrm{Res}_t(dw/w) = n$ according to ii).

Now we pass to the proof of prop. 5′. We write the differential form ω in the form

$$\omega = \sum_{n \geq 0} a_n\, du/u^n + \omega_0 \qquad \text{with } v(\omega_0) \geq 0.$$

Then $\mathrm{Res}_u(\omega) = a_1$ and $\mathrm{Res}_t(\omega) = \sum a_n\, \mathrm{Res}_t(du/u^n)$. As $\mathrm{Res}_t(du/u) = 1$ according to iv), everything finally comes down to proving the following formula:

v) $\mathrm{Res}_t(du/u^n) = 0 \qquad$ for $n \geq 2$.

When the field k has characteristic zero, $du/u^n = dg$, with

$$g = -1/(n-1)u^{n-1},$$

and formula v) is a simple consequence of formula iii). This argument no longer applies in characteristic $p > 0$, for we could have $n - 1 \equiv 0 \bmod p$. However, the case of characteristic p can be reduced to that of characteristic zero in the following way:

First, we can suppose, after multiplying u by a scalar factor, that

$$u = t + a_2 t^2 + a_3 t^3 + \cdots = t(1 + a_2 t + a_3 t^2 + \cdots).$$

We deduce that

$$\frac{1}{u^n} = \frac{1}{t^n}(1 - n a_2 t + \cdots + b_i t^i + \cdots),$$

where the b_i are polynomials in $a_2, \ldots, a_{i+1}$ with coefficients in $\mathbf{Z}$ and are independent of the characteristic (the integer n being fixed).

By multiplying with $du = dt + 2a_2t\,dt + \cdots + ia_i t^{i-1}\,dt + \cdots$, we deduce that

$$\frac{du}{u^n} = \frac{dt}{t^n} \cdot \sum_{i=0}^{\infty} c_i t^i,$$

where the c_i are, as before, polynomials in $a_2, \ldots, a_{i+1}$ with coefficients in $\mathbf{Z}$ and are independent of the characteristic.

In particular, $c_{n-1} = \mathrm{Res}_t\left(\frac{du}{u^n}\right)$. Since formula v) is valid in characteristic 0, the polynomial $c_{n-1}(a_2, \ldots, a_n)$ vanishes each time that its arguments a_i are taken in a field of characteristic zero. By virtue of the *principle of prolongation of algebraic identities* (Bourbaki, *Algèbre*, chap. IV, §2, no. 5), this polynomial is thus identically zero, which proves v) in the general case, and finishes the proof of prop. 5′.   □

Remark. It would be easy to replace the recourse to the principle of prolongation of algebraic identities with a "functorial" argument. One introduces, for each commutative ring A, the algebra $K_A = A((t))$ and its module of differentials $D'_A(K_A)$. There is a homomorphism $\mathrm{Res}_t : D'_A(K_A) \to A$ commuting with homomorphisms $A \to B$. One then proves the formula v) for $u = t + \sum_{i \geq 2} a_i t^i$ in three steps:

a) for A a field of characteristic 0 (by the method of the text).
b) for A an integral domain of characteristic 0 (by embedding A in its field of fractions and using a)).
c) for arbitrary A (by writing A as the quotient of a polynomial ring over $\mathbf{Z}$ and applying b) to this ring).

We leave the details of this proof to the reader.

12. Proof of the residue formula

We begin by checking the formula in a particular case:

Lemma 3. *The residue formula is true when the curve X is the projective line* $\mathbf{P}_1(k)$.

PROOF. In this case, the identity map $X \to X$ is a function t on X, and $k(X) = k(t)$. Every differential ω on X can be written $\omega = f(t)\,dt$, where $f(t)$ is a rational function of t. Decomposing f into simple elements (Bourbaki, *Algèbre*, chap. VII, §2, no. 3), we can suppose that $f = t^n$ or $f = 1/(t-a)^n$.

In the first case, the only pole of ω is the point at infinity and putting $u = 1/t$, we have $\omega = -du/u^{n+2}$. Thus $\mathrm{Res}_\infty(\omega) = 0$ and the sum of the residues is indeed zero.

In the second case, if $n = 1$, $\omega = dt/(t-a)$ has poles at a and ∞, with residues 1 and -1 respectively; if $n \geq 2$, the point a is the only pole, with a zero residue.

Thus we have checked the residue formula in all cases.   □

Now let X be any curve. We choose a function φ on X which is not constant. If X' denotes the projective line $\mathbf{P}_1(k)$, we can consider φ as a map $X \to X'$ which is evidently surjective; it makes X a "covering" of X', possibly ramified. Putting $E = k(X')$ and $F = k(X)$, the map φ defines an embedding of E in F; the field E is thus identified with the field $k(\varphi)$ generated by φ. Since X has dimension 1, $[F : F^p] = p$; if F' denotes the largest separable extension of E contained in F, there thus exists an integer $n \geq 0$ such that $F' = F^{p^n}$. The extension F/E is *separable* if and only if $n = 0$, in other words if $\varphi \notin F^p$; we assume this from now on.

If f is an element of F, its *trace* in F/E is well defined; it is an element of E which we will write $\operatorname{Tr}_{F/E}(f)$. The operation of trace can be extended to differentials in the following way:

The injection $E \to F$ defines a homomorphism from $D_k(E)$ to $D_k(F)$; as $d\varphi$ is an E-basis of $D_k(E)$ and $\varphi \notin F^p$, this homomorphism is injective and extends to an isomorphism of $D_k(E) \otimes_E F$ with $D_k(F)$. On the other hand, $\operatorname{Tr}_{F/E} : F \to E$ is E-linear; applying this homomorphism to the second term of $D_k(E) \otimes_E F$, we finally deduce an E-linear map

$$\operatorname{Tr}_{F/E} : D_k(F) \to D_k(E).$$

We can make this more explicit as follows: if ω is a differential on X, we write $\omega = f\, d\varphi$ and then

$$\operatorname{Tr}_{F/E}(\omega) = (\operatorname{Tr}_{F/E}(f))d\varphi.$$

Thus, to every differential ω on X we have associated a differential $\operatorname{Tr}(\omega)$ on $X' = \mathbf{P}_1(k)$. This operation enjoys the following property:

Lemma 4. *For every point $P \in X'$,*

$$\sum_{Q \to P} \operatorname{Res}_Q(\omega) = \operatorname{Res}_P(\operatorname{Tr}(\omega)),$$

the sum being over all the points $Q \in X$ such that $\varphi(Q) = P$.

Lemmas 3 and 4 imply the residue formula. Indeed, if ω is a differential on X, lemma 4 shows that

$$\sum_{Q \in X^{\cdot}} \operatorname{Res}_Q(\omega) = \sum_{P \in X'} \operatorname{Res}_P(\omega'),$$

with $\omega' = \operatorname{Tr}(\omega)$ and lemma 3 shows that this last sum is zero.

Thus it remains to prove lemma 4. It is a "semi-local" statement, i.e., local on X' but not on X. We are going to begin by reducing it to a purely local claim.

Let $\widehat{E}_P$ be the completion of E for the valuation v_P, and similarly let $\widehat{F}_Q$ be the completions of F for the valuations v_Q associated to the points Q

mapping to P. The v_Q "extend" v_P in the following sense: there exist integers $e_Q \geq 1$ such that $v_Q = e_Q v_P$ on E; conversely, one sees immediately that every valuation of F which extends v_P coincides with one of the v_Q. This is a typical situation of the "decomposition" of a valuation; the $\widehat{F}_Q$ are extensions of $\widehat{E}_P$ of degrees e_Q, and there is a canonical isomorphism (cf. for example [15], p. 60)

$$F \otimes_E \widehat{E}_P = \prod_{Q \to P} \widehat{F}_Q.$$

A trace formula follows immediately from this isomorphism:

$$\mathrm{Tr}_{F/E}(f) = \sum_{Q \to P} \mathrm{Tr}_Q(f), \qquad f \in F,$$

where Tr_Q denotes the trace in the extension $\widehat{F}_Q/\widehat{E}_P$.

Whence, taking into account the additivity of the residue,

$$\mathrm{Res}_P(\mathrm{Tr}(f)\, d\varphi) = \sum_{Q \to P} \mathrm{Res}_P(\mathrm{Tr}_Q(f)\, d\varphi).$$

The last formula reduces lemma 4 to the following result:

Lemma 5. *For every $f \in \widehat{F}_Q$, $\mathrm{Res}_Q(f\, d\varphi) = \mathrm{Res}_P(\mathrm{Tr}_Q(f)\, d\varphi)$.*

The proof of this lemma will be the object of the following no.

Remarks. 1) The preceding reduction does not use at all the hypothesis that X' is a projective line; it gives a proof of lemma 4 which is valid for any separable covering $X \to X'$.

2) Following Hasse, we have deduced the residue formula from lemma 4. We mention that the converse is possible: from the residue formula (proved by transcendental methods, or by means of the Cartier operator, or by any other method), one easily deduces lemma 4. One can even extend it to *inseparable* coverings using a suitable definition of the trace of a differential (the definition used above no longer applies). We will come back to this in chap. III, no. 3.

13. Proof of lemma 5

As in proposition 5, the question is *local*. We have a field of formal power series K and a finite separable extension L of K. If t (resp. u) denotes a uniformizing parameter of L (resp. K), we want to establish the formula

$$\mathrm{Res}_t(f\, du) = \mathrm{Res}_u(\mathrm{Tr}(f)\, du) \qquad \text{for all } f \in L. \tag{*}$$

Moreover, we can restrict to the case where f is of the form t^n, with $n \in \mathbf{Z}$.

This being so, first suppose that *the characteristic of the field k is zero*. Denoting by e the degree of L/K, we have $v_L(u) = e$, which shows that $u = w^e$ with w a uniformizing parameter of L. Replacing t by w, *we can suppose that $u = t^e$*. Thus we are talking about a cyclic extension in which the computation of the trace presents no difficulties. We find

$$\text{Tr}(t^n) = \begin{cases} 0 \text{ if } n \not\equiv 0 \bmod e \\ eu^{n/e} \text{ if } n \equiv 0 \bmod e. \end{cases}$$

We deduce that

$$\text{Res}_u(\text{Tr}(t^n)\,du) = \begin{cases} 0 \text{ if } n \neq -e \\ e \text{ if } n = -e. \end{cases}$$

On the other hand,

$$\text{Res}_t(t^n\,du) = \text{Res}_t(et^{n+e-1}\,dt) = \begin{cases} 0 \text{ if } n \neq -e \\ e \text{ if } n = -e \end{cases}$$

and we indeed find the same result.

Now we pass to the general case. We can write

$$u = t^e + \sum_{i>e} a_i t^i, \tag{**}$$

and conversely such a formula defines a subfield $k((u))$ of $k((t))$ such that $[k((t)) : k((u))] = e$; the extension $k((t))/k((u))$ is separable if and only if $u \notin k((t^p))$.

Formula (**) makes evident the fact that $\{1, t, t^2, \ldots, t^{e-1}\}$ is a *basis* of $k((t))/k((u))$; for every $n \in \mathbf{Z}$, we can thus write

$$t^n.t^i = \sum_{j=0}^{j=e-1} b_{n,i,j}(u).t^j, \qquad 0 \leq i \leq e-1,$$

the $b_{n,i,j}(u)$ being formal series in u:

$$b_{n,i,j}(u) = \sum b_{n,i,j,k} u^k.$$

For fixed n, the $b_{n,i,j}(u)$ form a matrix which is nothing other than the matrix associated to t^n in the *regular representation of $k((t))/k((u))$*. By virtue of the definition of the trace, we thus have $\text{Tr}(t^n) = \sum_{i=0}^{i=e-1} b_{n,i,i}(u)$, and the residue $c_n = \text{Res}(\text{Tr}(t^n)\,du)$ is given by the formula

$$c_n = \sum_{i=0}^{i=e-1} b_{n,i,i,-1}.$$

On the other hand, one sees immediately that $\text{Res}(t^n\,du) = -na_{-n}$ (agreeing to replace a_e by 1 and a_i by 0 if $i < e$), and the formula to be proved is thus equivalent to

$$c_n = -na_{-n} \qquad \text{for all } n \in \mathbf{Z}. \tag{***}$$

But the preceeding computations can be done "universally", considering the a_i as indeterminants. It follows that the $b_{n,i,j,k}$ are *polynomials* in the a_i with coefficients in $\mathbf{Z}$ and are independent of the characteristic. The same is thus true of $c_n + na_{-n}$. According to what we have seen above, these polynomials vanish each time their arguments are taken in an algebraically closed field of characteristic 0; applying the principle of prolongation of algebraic identities, we deduce that this polynomial is identically zero which finishes the proof of lemma, and at the same time, that of the residue formula. $\qquad\qquad\qquad\qquad\qquad\qquad\qquad\qquad\qquad\qquad\qquad\qquad\qquad$ □

Bibliographic note

Among the numerous works which treat algebraic curves, we limit ourselves to mentioning those of Severi [79], Weyl [97], Chevalley [15], and Weil [88] which suffice to give an idea of the various points of view. The lessons of Severi are written in the style of Italian algebraic geometry; they contain many interesting results on linear series, projective embeddings, and automorphisms of algebraic curves. Weyl takes the point of view of "analytic geometry", which leaves the purely algebraic realm; in particular, he proves the uniformization theorem, as well as the fact that every compact Riemann surface is algebraic. One knows that this last result leads to the determination of the coverings of a curve with given ramification (Riemann existence theorem), a determination which algebraic methods have not yet obtained.

Severi and Weyl limit themselves to the classical case, where the base field is $\mathbf{C}$. With Chevalley and Weil, the base field is arbitrary. This is almost the only point in common of their works: that of Chevalley is written in the purely algebraic style (always fields, never curves), while Weil employs the more geometric language of the Foundations [87].

From any point of view, the central theorem is the Riemann-Roch theorem. The proof that we have given, using répartitions, was introduced by Weil in a letter addressed to Hasse [85]. It is rapid and has the advantage of translating easily into the language of sheaves, thus preparing the way for generalizations to varieties of any dimension (see chap. IV, for the case of surfaces). It is interesting to note that this proof figures in the work of Chevalley [15] already cited, but not in that of Weil [88].

As we have seen, the residue formula plays an essential role in identifying differentials with linear forms on répartitions (the "duality" theorem). The first proof of this formula (over a field of any characteristic) is due to Hasse [32]; it is essentially his proof that we have given. The work of Chevalley [15] contains another, rather indirect, but avoiding the difficult lemma 5 (see also Lang [51], chap. X, §5). There is another proof of this lemma in a note of Whaples [98]. At any rate, these various proofs are artificial. Here, as with many other questions (see in particular chap. IV), it seems that one can obtain a truly natural proof only by taking the point of view of Grothendieck's general "duality theorem" [28]; for this see Altman-Kleiman [105], Hartshorne [115] as well as Tate [124].

CHAPTER III

Maps From a Curve to a Commutative Group

This chapter contains the proof of the first theorem stated in chapter I: the existence of a modulus associated to a rational map from an algebraic curve to a commutative algebraic group.

The proof itself is given in §2. We have preceded it, in §1, with a general study of "local symbols", and we have given the value of these symbols in some particular cases. Finally, §3 contains a certain number of auxiliary results, more or less well known, but for which it is difficult to give satisfactory references.

§1. Local symbols

1. Definitions

Let X be an algebraic curve (satisfying the conditions of chapter II, whose notations we keep). If S is a finite subset of X, we call the assignment of an integer $n_P > 0$ for each point $P \in S$ a *modulus* supported on S. The modulus $\mathfrak{m}$ will often be identified with the effective divisor $\sum n_P P$.

If g is a rational function on X, we will write

$$g \equiv 1 \bmod \mathfrak{m}$$

if $v_P(1 - g) \geq n_P$ for every $P \in S$.

If the equality above is only verified at a point P, we will write

$$g \equiv 1 \bmod \mathfrak{m} \text{ at } P.$$

Note that, if $g \equiv 1 \bmod \mathfrak{m}$, the divisor (g) of g is prime to S.

Now let $f : X - S \to G$ be a map from the complement of S to a commutative group G. (Note that *we do not suppose* that G is an algebraic group, nor, if it is, that f is a rational map.) The map f extends by linearity to a homomorphism from the group of divisors prime to S to the group G. In particular, if $g \equiv 1 \bmod \mathfrak{m}$, the element $f((g)) \in G$ is well defined and writing the group G additively we have

$$f((g)) = \sum_{P \in X - S} v_P(g) f(P).$$

Definition 1.   We say that $\mathfrak{m}$ is a modulus for the map f (or that $\mathfrak{m}$ is associated to f) if $f((g)) = 0$ for every function $g \in k(X)$ such that $g \equiv 1 \bmod \mathfrak{m}$.

We are going to transform this definition using the notion of a "local symbol."

Definition 2.   Let $\mathfrak{m}$ be a modulus supported on S and let f be a map from $X - S$ to G. We will call a "local symbol" the assignment, for each $P \in X$ and every $g \in k(X)^*$, of an element of G, written $(f, g)_P$, satisfying the following four conditions:

i) $(f, gg')_P = (f, g)_P + (f, g')_P$.
ii) $(f, g)_P = 0$ if $P \in S$ and if $g \equiv 1 \bmod \mathfrak{m}$ at P.
iii) $(f, g)_P = v_P(g) f(P)$ if $P \in X - S$.
iv) $\sum_{P \in X} (f, g)_P = 0$.

Some examples of local symbols will be given in nos. 3 and 4.

Proposition 1.   *In order that $\mathfrak{m}$ be a modulus for the map f, it is necessary and sufficient that there exist a local symbol associated to f and to $\mathfrak{m}$, and this symbol is then unique.*

PROOF.   Suppose that a local symbol exists and let g be a function such that $g \equiv 1 \bmod \mathfrak{m}$; then

$$f((g)) = \sum_{P \notin S} v_P(g) f(P)$$

$$= \sum_{P \notin S} (f, g)_P \qquad\qquad \text{using iii)}$$

$$= - \sum_{P \in S} (f, g)_P \qquad\qquad \text{using iv)}$$

$$= 0 \qquad\qquad\qquad\qquad \text{using ii)}$$

Conversely, suppose that $\mathfrak{m}$ is a modulus for f; we seek to define a local symbol $(f, g)_P$. If $P \notin S$, condition iii) imposes $(f, g)_P = v_P(g) f(P)$. Thus suppose $P \in S$. One can always find an auxiliary function g_P such that $g_P \equiv 1$ mod $\mathfrak{m}$ at the points $Q \in S - P$, and such that $g/g_P \equiv 1$ mod $\mathfrak{m}$ at P (the existence of g_P follows, for example, from the approximation theorem for valuations). We then define $(f, g)_P$ by the formula

$$(f, g)_P = - \sum_{Q \notin S} v_Q(g_P) f(Q). \tag{*}$$

The right hand side does not depend on the auxiliary function g_P chosen; indeed, one can only change g_P by multiplying it by a function h such that $h \equiv 1$ mod $\mathfrak{m}$ and that does not change the sum in question, since $f((h)) = 0$.

The formula (*) thus defines $(f, g)_P$ unambiguously, when $P \in S$. It remains to see that the properties i), ii), iii), and iv) hold:

Verification of i): If g_P and g'_P are auxiliary functions for g and g' respectively, we can take $g_P g'_P$ as an auxiliary function for gg' and the formula follows immediately.

Verification of ii): If $g \equiv 1$ mod $\mathfrak{m}$ at P, then $g_P \equiv 1$ mod $\mathfrak{m}$ and the right hand side of (*) is equal to $-f((g_P)) = 0$, since $\mathfrak{m}$ is a modulus for f.

Verification of iii): This is the very definition of $(f, g)_P$ when $P \notin S$.

Verification of iv): We have

$$\sum_{P \in S} (f, g)_P = - \sum_{P \in S} \sum_{Q \notin S} v_Q(g_P) f(Q)$$

$$= - \sum_{Q \notin S} v_Q(h) f(Q), \quad \text{with} \quad h = \prod_{P \in S} g_P.$$

Putting $g/h = k$, clearly $k \equiv 1$ mod $\mathfrak{m}$, whence

$$\sum_{Q \notin S} v_Q(k) f(Q) = 0$$

since $\mathfrak{m}$ is a modulus for f. This equality can thus be written

$$\sum_{P \in S} (f, g)_P = - \sum_{Q \notin S} v_Q(g) f(Q) + \sum_{Q \notin S} v_Q(k) f(Q)$$

$$= - \sum_{Q \notin S} v_Q(g) f(Q)$$

$$= - \sum_{Q \notin S} (f, g)_Q \quad \text{using iii).}$$

Thus, the expression (*) is indeed a local symbol associated to f and $\mathfrak{m}$. Moreover, it is the only possible, for according to ii) we must have

$$(f, g)_P = (f, g_P)_P,$$

and according to ii), iii), and iv), $(f, g_P)_P$ must be equal to the right hand side of the formula (*). The proof of prop. 1 is thus finished. $\square$

Remark. If a map $f : X - S \to G$ has a modulus $\mathfrak{m}$, it has others (for example the moduli $\mathfrak{m}' \geq \mathfrak{m}$), but the corresponding local symbols are the same. Indeed, we can restrict to the case where $\mathfrak{m}' \geq \mathfrak{m}$, and in this case, a local symbol for $\mathfrak{m}$ is one for $\mathfrak{m}'$ thus coincides with that of $\mathfrak{m}'$, according to the uniqueness property that we have just proved. Thus, the local symbol, if it exists, *only depends on f*.

Interpretation in terms of idèles. The preceding can easily be translated to Chevalley's language of "idèles". We rapidly indicate how:

Let I be the group of idèles of X, i.e., the multiplicative group of invertible elements in the ring of répartitions (chap. II, no. 5). Write F for the field $k(X)$ and, for every $P \in X$, denote by U_P the subgroup of F^* formed by the functions g such that $v_P(g) = 0$; if $n \geq 1$, denote by $U_P^{(n)}$ the subgroup of U_P formed by the functions such that $v_P(1 - g) \geq n$. With these notations, an idèle a is nothing other than a family $\{a_P\}_{P \in X}$ of elements of F^* such that $a_P \in U_P$ for almost all P.

Given this, let $(f, g)_P$ be a local symbol and put, for every idèle a

$$\theta(a) = \sum_{P \in X} (f, a_P)_P \quad \text{if } a = \{a_P\}_{P \in X}.$$

From the fact that $a_P \in U_P$ for almost all P, this sum is indeed finite, and thus we get a homomorphism $\theta : I \to G$. Moreover, the knowledge of this homomorphism *is equivalent to* to the knowledge of the local symbol $(f, g)_P$ with which we started. Conditions ii) and iii) imply that θ is zero on the subgroup $I_\mathfrak{m}$ of I defined by the formula

$$I_\mathfrak{m} = \prod_{P \in S} U_P^{(n_P)} \times \prod_{P \notin S} U_P \quad \text{if } \mathfrak{m} = \sum n_P P.$$

As for condition iv), it says that θ vanishes on the subgroup F^* of I formed by the principal idèles. Thus, θ is a homomorphism from $I/I_\mathfrak{m} F^*$ to G (and conversely, every homomorphism from $I/I_\mathfrak{m} F^*$ to G can be obtained in this way). It is easy to see, using the approximation theorem, that the group $I/I_\mathfrak{m} F^*$ is canonically isomorphic to the group $C_\mathfrak{m}$ introduced in chap. I, no. 1; this is essentially the content of prop. 1.

2. First properties of local symbols

a) *Functorial character* Let $f : X - S \to G$ be a map from $X - S$ to a commutative group G and let $\theta : G \to G'$ be a homomorphism from G to a commutative group G'. Then we have:

Proposition 2. *If $\mathfrak{m}$ is a modulus for f, it is also a modulus for $\theta \circ f$, and the corresponding local symbols satisfy the formula*

$$(\theta \circ f, g)_P = \theta((f, g)_P).$$

PROOF. It suffices to check that $\theta((f,g)_P)$ is a local symbol associated to $\theta \circ f$ and to $\mathfrak{m}$, in other words that the properties i), ii), iii), and iv) are satisfied. This is immediate.   □

b) *Local symbol of a trace.* Again let $f : X - S \to G$ and suppose that $\pi : X \to X'$ is a map from X onto another curve X' (we can thus consider X as a "ramified covering" of X', cf. chap. II, no. 12). Put $S' = \pi(S)$ and, for every $P' \in X'$, denote by $\pi^{-1}(P')$ the divisor of X which is the inverse image of P' by π. We have

$$\pi^{-1}(P') = \sum_{P \to P'} e_P P,$$

where e_P denotes the index of ramification of the valuation v_P with respect to the valuation $v_{P'}$.

If $P' \notin S'$, the divisor $\pi^{-1}(P')$ is prime to S, and $f(\pi^{-1}(P'))$ makes sense. Thus we get a map

$$\mathrm{Tr}_\pi \, f : X' - S' \to G$$

which will be called the *trace* of the map f.

Proposition 3. *If f has a modulus $\mathfrak{m}$, the map $f' = \mathrm{Tr}_\pi \, f$ has a modulus $\mathfrak{m}'$ and*

$$(\mathrm{Tr}_\pi \, f, g')_{P'} = \sum_{P \to P'} (f, g' \circ \pi)_P \quad P' \in X', \quad g' \in k(X')^*.$$

PROOF. We must see that the expression $(f', g')_{P'}$ defined above satisfies the conditions i), ii), iii), and iv) for f' and a suitable modulus $\mathfrak{m}$. For i), this is evident. For ii), put

$$\mathfrak{m} = \sum_{P \in S} n_P P$$

and, for every $P' \in S'$, choose an integer $n_{P'}$ which is larger than all the quotients n_P / e_P for $P \in S \cap \pi^{-1}(P')$. If $v_{P'}(1 - g') \geq n_{P'}$, we deduce that

$$v_P(1 - g' \circ \pi) \geq e_P n_{P'} \geq n_P \quad \text{if } P \to P' \text{ and } P \in S,$$

whence $(f, g' \circ \pi)_P = 0$ in this case. If $P \notin S$, the fact that $v_P(g' \circ \pi) = 0$ implies $(f, g' \circ \pi)_P = 0$; condition ii) is thus satisfied by the modulus

$$\mathfrak{m}' = \sum_{P' \in S'} n_{P'} P'.$$

For iii), we must compute $(\mathrm{Tr}_\pi\, f, g')_{P'}$ for $P' \notin S'$. We then have

$$(\mathrm{Tr}_\pi\, f, g')_{P'} = \sum_{P \to P'} v_P(g' \circ \pi) f(P) \quad \text{(since } P \notin S),$$

$$= v_{P'}(g') \sum_{P \to P'} e_P f(P)$$

$$= v_{P'}(g') f'(P'), \qquad \text{by the very definition of } f'.$$

Condition iii) is thus fulfilled, and the same is evidently true of condition iv). $\qquad\qquad\qquad\qquad\qquad\qquad\qquad\qquad\qquad\qquad\qquad\qquad\qquad\qquad$ $\square$

c) *Local symbol of a norm.* Let $\pi : X \to X'$ be a ramified covering, let S' be a finite subset of X', and let $f' : X' - S' \to G$ be a map from $X' - S'$ to a commutative group G. We put $S = \pi^{-1}(S')$. On the other hand, if $g \in k(X)^*$, we denote by $N_\pi g$ the norm of g in the extension $k(X)/k(X')$ defined by π. We then have the following proposition, analogous to proposition 3, but where the roles of f and g have been permuted:

Proposition 4. *If f' has a modulus $\mathfrak{m}'$, the map $f' \circ \pi$ has a modulus $\mathfrak{m}$, and we have*

$$(f', N_\pi g)_{P'} = \sum_{P \to P'} (f' \circ \pi, g)_P \quad P' \in X', \; g \in k(X)^*.$$

PROOF. First we observe the existence of a modulus $\mathfrak{m}$ with support S such that $g \equiv 1 \bmod \mathfrak{m}$ implies $N_\pi g \equiv 1 \bmod \mathfrak{m}'$: indeed this is a well-known result on norms (the norm map is "continuous") that one can prove, for example, by embedding the extension $k(X)/k(X')$ in a normal extension. On the other hand, if $g \in k(X)^*$, one knows that $(N_\pi g) = \pi((g))$. Applying this to the case $g \equiv 1 \bmod \mathfrak{m}$, we see that $f' \circ \pi((g)) = f'(N_\pi g) = 0$, which indeed shows that $\mathfrak{m}$ is a modulus for $f' \circ \pi$.

It remains to establish the formula linking the local symbols of f' and $f' \circ \pi$. If $P' \in X' - S'$ this formula simply says $v_{P'}(N_\pi g) = \sum_{P \to P'} v_P(g)$, i.e., that $(N_\pi g) = \pi((g))$. Thus suppose $P' \in S'$, and choose a function h such that $g/h \equiv 1 \bmod \mathfrak{m}$ at the points P mapping to P' and $h \equiv 1 \bmod \mathfrak{m}$ at the points $P \in S$ not mapping to P'. Then $N_\pi g/N_\pi h \equiv 1 \bmod \mathfrak{m}'$ at P' and $N_\pi h \equiv 1 \bmod \mathfrak{m}'$ on $S' - P'$. Whence

$$(f', N_\pi g)_{P'} = (f', N_\pi h)_{P'} = - \sum_{Q' \notin S'} (f', N_\pi h)_{Q'}$$

$$= - \sum_{P \notin S} (f' \circ \pi, h)_P$$

$$= \sum_{P \to P'} (f' \circ \pi, h)_P$$

$$= \sum_{P \to P'} (f' \circ \pi, g)_P. \qquad\qquad\qquad \square$$

3. Example of a local symbol: additive group case

From now on, we limit ourselves to the case where the commutative group G is a connected *algebraic* group, the map $f : X - S \to G$ being a *regular* map. We can then consider f as a *rational* map from X to G, regular away from S. Unless otherwise stated, we suppose that S is the *smallest* subset of X having this property, in other words it is the set of points where f is not regular.

The theorem of Rosenlicht that we propose to prove in the rest of this chapter can be stated thus (cf. chap. I, thm. 1):

Theorem 1. *The map f has a modulus supported on S.*

We are going to verify this theorem in the particular case where the group G is the additive group $\mathbf{G}_a$ and at the same time complete it by determining explicitly the local symbol $(f, g)_P$.

Proposition 5. *Theorem 1 is true for the group $\mathbf{G}_a$, the corresponding local symbol being $(f, g)_P = \mathrm{Res}_P(f \, dg/g)$.*

(This formula makes sense, for f is nothing other than a (scalar) function on X with S as its set of poles.)

PROOF. If P belongs to S, we put $n_P = 1 - v_P(f)$; from the fact that P is a pole of f, we have $n_P > 1$. We are going to check that $\mathrm{Res}_P(f \, dg/g)$ is a local symbol associated to f and $\mathfrak{m} = \sum n_P P$.

Property i) is clear, from the fact that

$$d(gg')/gg' = dg/g + dg'/g'.$$

For ii), we remark that, if $v_P(1 - g) \geq n_P$, then

$$v_P(dg) \geq n_P - 1 \geq -v_P(f);$$

as $v_P(g) = 0$ we deduce that $v_P(f \, dg/g) \geq 0$, whence $\mathrm{Res}_P(f \, dg/g) = 0$.

For iii), we remark that dg/g has a *simple* pole at P, thus so does $f \, dg/g$ (since $P \notin S$) and we have

$$\mathrm{Res}_P(f \, dg/g) = f(P) \, \mathrm{Res}_P(dg/g) = f(P) v_P(g),$$

applying formula iv) of chap. II, no. 11.

Finally, formula iv):

$$\sum_{P \in X} \mathrm{Res}_P(f \, dg/g) = 0$$

is just the residue formula applied to the differential form $\omega = f \, dg/g$.

The proof of prop. 5 is thus complete.   □

Corollary. *In characteristic $p > 0$, we have the formula*

$$\operatorname{Res}_P(f^p\, dg/g) = [\operatorname{Res}_P(f\, dg/g)]^p.$$

PROOF. Indeed, the map $x \to x^p$ is a homomorphism $\mathbf{G}_a \to \mathbf{G}_a$ and we know that local symbols are functorial (prop. 2).   □

(Of course, nothing could be easier than to check this formula by a direct computation!)

We could also apply prop. 3 to the present case. We would recover the formula giving the residue of a trace (chap. II, no. 12, lemma 4):

$$\sum_{P \to P'} \operatorname{Res}_P(\omega) = \operatorname{Res}_{P'}(\operatorname{Tr} \omega).$$

Proposition 4 also gives this formula, taking into account the formula $d(N_\pi g)/N_\pi g = \operatorname{Tr}_\pi(dg/g)$.

Remark. Proposition 5 extends to the *Witt group W_n* of any dimension n. A rational map from X to W_n is nothing other than a Witt vector $\vec{f}$ of length n with components in $k(X)$. One should take for $(\vec{f}, g)_P$ the symbol defined by Witt ([99], §2), and use a formula similar to the residue formula (loc. cit., §9). See also Kawada and Satake [43].

4. Example of a local symbol: multiplicative group case

Suppose now that G is the multiplicative group $\mathbf{G}_m$. The rational map f can again be identified with a function on X and S consists of the set of zeroes and poles of f.

Proposition 6. *The map f has $\mathfrak{m} = \sum_{P \in S} P$ as a modulus; the corresponding local symbol is*

$$(f, g)_P = (-1)^{mn} \frac{f^n}{g^m}(P), \quad \text{with } n = v_P(g), \ m = v_P(f).$$

(This formula makes sense, for the function $h = f^n/g^m$ is such that $v_P(h) = 0$, thus it has a well-defined, non-zero value at the point P.)

PROOF. Here again we must check the properties i), ii), iii), and iv) of a local symbol.

For i), let $g'' = gg'$. Then $n'' = n + n'$, whence

$$(f, g'')_P = (-1)^{(n+n')m} \frac{f^{n+n'}}{g^m g'^m}(P) = (f, g)_P (f, g')_P.$$

For ii), suppose that $v_P(1 - g) \geq 1$; then $n = 0$, whence

$$(f, g)_P = \frac{1}{g^m}(P) = 1 \quad \text{since} \quad g(P) = 1.$$

For iii), suppose that $P \notin S$, i.e., that $m = v_P(f) = 0$. Then $(f,g)_P = f(P)^{v_P(g)}$.

It remains only to check the following formula ("product formula"):

iv)
$$\prod_{P \in X} (f,g)_P = 1.$$

We are going to proceed as for the residue formula, reducing to the case where X is the projective line Λ. The function g can be considered as a map $g : X \to \Lambda$. If g is a constant map equal to a, the left hand side of iv) is equal to a raised to the power $-\sum v_P(f) = -\deg((f)) = 0$, and the formula iv) is correct in this case. We can thus suppose that g is not constant, in which case it is a surjective map, which makes X a (ramified) covering of Λ. Putting $F = k(X)$ and $E = k(\Lambda)$, we thus have an extension F/E with $E = k(g)$, and the *norm* operation $N_{F/E} : F^* \to E^*$ is well defined. Then denoting by t the identity map $\Lambda \to \Lambda$, considered as an element of $k(\Lambda)$, we are going to establish the following two lemmas:

Lemma 1. *For every point* $P \in \Lambda$, $\prod_{g(Q)=P}(f,g)_P = (N_{F/E}\,f, t)_P$.

Lemma 2. *For every function* f' *on* Λ, $\prod_{P \in \Lambda}(f', t)_P = 1$.

It is clear that the formula iv) follows from lemma 1 and lemma 2 applied to $f' = N_{F/E}\,f$. It thus remains to prove these two lemmas.

PROOF OF LEMMA 2. We write f' in the form $f' = \mu \prod (t - \lambda)^{n_\lambda}$. As the symbol $(f', t)_P$ is multiplicative in f', we can restrict to $f' = t - \lambda$. There are two cases to distinguish: $\lambda = 0$ and $\lambda \neq 0$:

a) $\lambda = 0$. Then $(t,t)_P = 1$ for $P \neq 0, \infty$, $(t,t)_0 = -1$, $(t,t)_\infty = -1$, and the product is indeed equal to 1.

b) $\lambda \neq 0$. Then $(t - \lambda, t)_P = 1$ for $P \neq 0, \lambda, \infty$;

$$(t - \lambda, t)_P = -\lambda$$
$$(t - \lambda, t)_P = 1/\lambda$$
$$(t - \lambda, t)_P = -1$$

and the product is indeed again equal to 1.   □

PROOF OF LEMMA 1. We are going to reduce it to a local result, as we did in chap. II, no. 12 for lemma 4. First of all, we observe that the symbol $(f', t)_P$ makes sense when f' and t are any elements of the field $K = \widehat{E}_P$, the completion of the field E with repect to the valuation v_P. The symbol thus obtained will be written $(f', t)_K$; it is again multiplicative in f' and in t. Proceeding similarly with the local field $L = \widehat{F}_Q$, we have the formula

$$N_{F/E}\,f = \prod_{Q \to P} N_Q\,f, \quad \text{with} \quad N_Q = N_{\widehat{F}_Q/\widehat{E}_P}.$$

The right hand side of the formula to prove can thus be written

$$\prod_{Q \to P} (N_Q \, f, t)_K,$$

and we are reduced to proving the following result:

Lemma 3. $(f,g)_L = (N_Q \, f,g)_K$ if $f \in L^*$, $g \in K^*$.

(In this statement, the field K is identified with a subfield of L, which has the effect of identifying t with g.)

PROOF. It suffices to do the proof when f is a uniformiser of the field L; indeed, $(f,g)_L$ and $(N \, f,g)_K$ are both multiplicative in f, and the group L^* is generated by elements which are uniformisers (since every "unit" is the quotient of two uniformisers). We thus have $L = k((f))$. Similarly, we can suppose that g is a uniformiser of K, whence $K = k((g))$. If we then compute $v_P(N \, f) = v_Q(f)$, we find 1, which shows that

$$(N \, f,g)_K = -\frac{N \, f}{g}(P).$$

On the other hand, putting $e = [L : K]$ we have $v_Q(g) = e$, whence

$$(f,g)_L = (-1)^e \frac{f^e}{g}(Q).$$

Everything thus comes down to showing that the function $N \, f/f^e$ takes the value $(-1)^{e-1}$ at the point P (or at the point Q, it is the same thing). For this, we write the minimal equation of f over K

$$f^e + a_1 f^{e-1} + \cdots + a_e = 0, \qquad a_i \in K.$$

Clearly $a_e = (-1)^e N \, f$. On the other hand, if $a_i \neq 0$,

$$v_L(a_i f^{e-i}) = e.v_K(a_i) + e - i \equiv -i \bmod e.$$

It follows that all the monomials of the preceding equation have distinct valuations, except perhaps f^e and a_e. Elementary properties of valuations then show that we must have

$$v_L(f^e) = v_L(a_e) > v_L(a_i f^{e-i}) \quad \text{if} \quad 1 \le i \le e - 1.$$

Dividing by f^e, this gives $v_L(1 + a_e/f^e) > 0$, which means that a_e/f^e takes the value -1 at P; thus $N \, f/f^e$ takes the value $(-1)^{e-1}$, as was to be shown. $\qquad\square$

Remark. As with the Hilbert norm residue symbol $(\frac{\alpha,\beta}{p})$, the symbol $(f,g)_P$ has the following properties (which one checks by a direct computation):

$$\begin{cases} (f,g)_P(g,f)_P = 1 \\ (-f,f)_P = 1 \\ (1-f,f)_P = 1. \end{cases}$$

We also mention the following result:

Proposition 7. *If f and g are two functions on X whose divisors are disjoint, $f((g)) = g((f))$.*

PROOF. Write $\prod_{P \in X}(f, g)_P = 1$, taking into account that $(f, g)_P$ is equal either to $f(P)^{v_P(g)}$, or to $g(P)^{-v_P(f)}$; thus $f((g))g(-(f)) = 1$, whence the desired result. □

§2. Proof of theorem 1

5. First reduction

We return to the situation of no. 2 b), and suppose given a covering π : $X \to X'$. If $f : X - S \to G$ is any map from $X - S$ to a commutative group G, we have defined $\text{Tr}_\pi f : X' - S' \to G$ where $S' = \pi(S)$.

Proposition 8. *If G is an algebraic group and if f is a regular map, the map $\text{Tr}_\pi f$ is regular.*

PROOF. Let n be the degree of the covering $X \to X'$, and let $X^{(n)}$ be the *n-fold symmetric product* of X (cf. no. 14). For every $P' \in X'$ the divisor

$$\pi^{-1}(P') = \sum_{P \to P'} e_P P \qquad \text{(cf. no. 2)}$$

is an effective divisor of degree n, thus can be identified with a point of $X^{(n)}$. The map $\pi^{-1} : X' \to X^{(n)}$ thus defined is a *regular* map (cf. no. 15, proposition 22). On the other hand put $Y = X - S$; it is an open subset of X. Putting

$$F(y_1, \ldots, y_n) = f(y_1) + \cdots + f(y_n)$$

we get a regular map of Y^n to G which is invariant under permutation of the y_i. In view of the definition of the symmetric product, this map passes to the quotient and defines $F' : Y^{(n)} \to G$. But $Y^{(n)}$ is an open of $X^{(n)}$ and π^{-1} maps $Y' = X' - S'$ to $Y^{(n)}$; it follows that $F' \circ \pi^{-1} : X' - S' \to G$ is a regular map. As this map is nothing other than $\text{Tr}_\pi f$, this proves the proposition. □

The proposition applies in particular to the case of a non-constant rational function g on X, g being considered as a map from X to the projective line Λ. If $\mathfrak{m} = \sum_{P \in S} n_P P$, $n_P > 0$, is a modulus supported on S, we will write $g \equiv 0 \bmod \mathfrak{m}$ if $v_P(g) \geq n_P$ for all $P \in S$. Under these conditions,

$g(P) = 0$ for all $P \in S$, and the set $S' = g(S)$ is thus reduced to the point $\{0\}$ of Λ. The map $\text{Tr}_g\, f$, by virtue of the preceding proposition, is thus *a regular map from $\Lambda - \{0\}$ to G.*

Proposition 9. *Let $\mathfrak{m} = \sum_{P \in S} n_P P$, $n_P > 0$ be a modulus supported on S. In order that $\mathfrak{m}$ be a modulus for f, it is necessary and sufficient that for every non-constant function g with $g \equiv 0 \bmod \mathfrak{m}$, the map $\text{Tr}_g\, f$ be constant.*

PROOF. Put $f' = \text{Tr}_g\, f$ where g is a non-constant function such that $g \equiv 0 \bmod \mathfrak{m}$. If a is a point of Λ distinct from 0, then $f'(a) = f(g^{-1}(a))$ by the definition of f'. If $a = \infty$, the divisor $g^{-1}(\infty)$ is nothing other than the divisor $(g)_\infty$ of poles of the function g; if $a \neq \infty$, it is the divisor $(g)_a$ of zeros of the function $g - a$. Thus, in every case

$$f'(a) = f((g)_a) \qquad a \in \Lambda - \{0\}.$$

Now suppose that $\mathfrak{m}$ is a modulus for f. For every $a \neq 0, \infty$, we can write $g - a = -a(1 - a^{-1}g) = b.h$, where b is a non-zero constant and $h \equiv 1 \bmod \mathfrak{m}$. It follows, in view of the hypothesis made on $\mathfrak{m}$, that we have $f((g - a)) = 0$, which can be written

$$f((g)_a) = f((g)_\infty),$$

or

$$f'(a) = f'(\infty),$$

and this shows that f' is a constant map.

Conversely, suppose that f' is a constant map for every function $g \equiv 0 \bmod \mathfrak{m}$, g non-constant. If $h \equiv 1 \bmod \mathfrak{m}$, we can write $h = 1 - g$ with $g \equiv 0 \bmod \mathfrak{m}$. If g is constant, the same is true of h, and $f((h)) = f(0) = 0$. If g is not constant, $(h) = (g)_1 - (g)_\infty$, whence $f((h)) = f'(1) - f'(\infty) = 0$, as was to be shown. $\square$

6. Proof in characteristic 0

Let, as before, $f : X - S \to G$ be a regular map from $X - S$ to the commutative algebraic group G. Denote by r the dimension of G and let $\{\omega_1, \ldots, \omega_r\}$ be a basis of the vector space of differential forms of degree 1 *invariant by translation* on G (for the properties of these forms, cf. no. 11). Put $\alpha_i = f^*(\omega_i)$ $1 \leq i \leq r$; the α_i are the pull-backs of the ω_i by f, and are thus regular on $X - S$. For $P \in S$, we choose an integer $n_P > 0$ such that $v_P(\alpha_i) \geq -n_P$ for $1 \leq i \leq r$.

Proposition 10. *In characteristic 0, the modulus $\mathfrak{m} = \sum_{P \in S} n_P P$ defined above is a modulus for f.*

PROOF. We are going to apply the criterion of prop. 9. Thus let g be a non-constant rational function on X such that $g \equiv 0 \bmod \mathfrak{m}$. We are going to show that the map $\operatorname{Tr}_g f : \Lambda - \{0\} \to G$ is a constant map. Denote this map by f'. *It will suffice to prove that* $f'^*(\omega_i) = 0$ *for* $1 \le i \le r$. Indeed, from the fact that the ω_i are linearly independent at each point of G, this implies that the tangent map to f' is everywhere zero, whence the fact that f' is constant since the characteristic is zero.

First of all we have:

Lemma 4. $f'^*(\omega_i) = \operatorname{Tr}(\alpha_i)$ *for* $1 \le i \le r$.

(The trace is taken with respect to the covering $g : X \to \Lambda$, cf. chap. II, no. 12.)

PROOF. Let Y be a Galois covering dominating X (cf. no. 13) and let π be the projection $Y \to X$. From the fact that $g \circ \pi : Y \to \Lambda$ is a *separable* covering, the map

$$\omega \to (g \circ \pi)^*(\omega) = \pi^* g^*(\omega)$$

is an *injective* map from the differential forms of Λ to those of Y. It thus will suffice to prove the formula

$$\pi^* g^* f'^*(\omega_i) = \pi^* g^* \operatorname{Tr}(\alpha_i). \qquad (*)$$

The right hand side of (*) can be written

$$\sum_{j=1}^{j=n} \sigma_j^* \pi^*(\alpha_i),$$

the σ_j being certain elements of the Galois group $\mathfrak{g}$ of the covering $Y \to \Lambda$ (cf. no. 13). On the other hand, we also have

$$f' \circ g \circ \pi = \sum_{j=1}^{j=n} f \circ \pi \circ \sigma_j \qquad \text{(cf. no. 13)};$$

by virtue of the additivity of the operation $f^*(\omega)$ (no. 11, proposition 17), the left hand side of (*) can be written

$$\pi^* g^* f'^*(\omega_i) = \sum_{j=1}^{j=n} \sigma_j^* \pi^* f^*(\omega_i) = \sum_{j=1}^{j=n} \sigma_j^* \pi^*(\alpha_i),$$

and we do find the same result. $\qquad\qquad\qquad\qquad\qquad\qquad \square$

(Observe that this proof is valid in any characteristic, provided that g is not a p-th power. Indeed, it is likely that lemma 4 is valid without hypothesis on g, cf. Barsotti [6], thm. 4.2.)

Lemma 5. *For every i, $1 \leq i \leq r$, the differential form $f'^*(\omega_i)$ has at most a simple pole at 0.*

PROOF. We must prove that $v_0(f'^*(\omega_i)) \geq -1$. Thus suppose that $v_0(f'^*(\omega_i)) = -m - 1$ with $m \geq 1$. If $t : \Lambda \to \Lambda$ denotes the identity map considered as a rational function on Λ, then

$$v_0(t^m f'^*(\omega_i)) = -1, \quad \text{whence} \quad \text{Res}_0(t^m f'^*(\omega_i)) \neq 0.$$

According to lemma 4, the differential form $t^m f'^*(\omega_i)$ is the trace of $g^m \alpha_i$. Applying lemma 4 of chap. II, no. 12, we thus get

$$\sum_{P \to 0} \text{Res}_P(g^m \alpha_i) \neq 0, \qquad 1 \leq i \leq r.$$

But the form $g^m \alpha_i$ is regular at all the points P such that $g(P) = 0$. Indeed, either $P \notin S$, and then g and α_i are regular at this point or $P \in S$ and

$$v_P(g^m \alpha_i) = m v_P(g) + v_P(\alpha_i) \geq m n_P - n_P \geq 0.$$

The hypothesis $m \geq 1$ thus leads to a contradiction, which establishes lemma 5. □

It is now easy to finish the proof of proposition 10. Indeed, since f' is regular on $\Lambda - \{0\}$, the same is true of $f'^*(\omega_i)$; by virtue of lemma 5, this differential has at most a simple pole at 0. The residue formula then shows that its residue at this point is 0 and $f'^*(\omega_i)$ is a differential of the first kind, thus it is zero because Λ is a curve of genus zero. This finishes the proof, taking into account what was said above. □

7. Proof in characteristic $p > 0$: reduction of the problem

Our proof will rely on the *structure* of commutative algebraic groups. We assume the following two results:

Proposition 11 ("Chevalley's theorem"). *Every connected algebraic group G contains a normal subgroup R such that*:

a) *R is a connected linear group.*
b) *G/R is an Abelian variety.*

(For the definition of quotient groups, see Chevalley [17], exposé 8.)

Proposition 12. *Every connected commutative linear algebraic group is isomorphic to the product of a certain number of multiplicative groups G_m and a group U isomorphic to a subgroup of the group of triangular matrices having only 1's on the principal diagonal.*

(Such a group U is called "unipotent".)

Let us apply prop. 11 to the group G. We thus have $G/R = A$, where A is an Abelian variety; on the other hand, prop. 12 permits us to decompose R as a direct product of a unipotent group and of groups $\mathbf{G}_m$. In general, suppose that $R = R_1 + R_2$; then the group G is embedded as a subgroup in $G/R_1 \times G/R_2$ and, if $\mathfrak{m}_i$ is a modulus for the composed map

$$X - S \xrightarrow{f} G \to G/R_i \qquad i = 1, 2.$$

it is clear that $\mathfrak{m} = \operatorname{Sup}(\mathfrak{m}_1, \mathfrak{m}_2)$ is a modulus for f. We are thus reduced to proving theorem 1 for the groups G/R_1 and G/R_2, which are extensions of A by R_1 and R_2. Proceeding closer and closer, *we are reduced to proving theorem 1 in the two following cases:*

a) *G is an extension of an Abelian variety A by a group $\mathbf{G}_m$.*
b) *G is an extension of an Abelian variety A by a unipotent group U.*

We will prove a) in the next no., and b) in nos. 9 and 10.

Remark. Thus Chevalley's theorem serves us in proving theorem 1. Conversely, as Rosenlicht noted, if one could prove theorem 1 directly, one would deduce a new proof of Chevalley's theorem via the theory of generalized Jacobians (cf. chap. VII, no. 13).

8. Proof in characteristic $p > 0$: case a)

We are going to need the following two elementary lemmas:

Lemma 6. *Every regular map from the projective line minus one point to the multiplicative group $\mathbf{G}_m$ is constant.*

PROOF. We can suppose that the point in question is the point at infinity. In this case, the map is a rational function which has neither poles nor zeroes at a finite distance; it is thus a polynomial without zeroes, i.e., a constant. $\square$

Lemma 7. *Let V be a non-singular variety and let Ω be the sheaf of differential forms of degree 1 (cf. chap. II, no. 7). Suppose that, for every $P \in V$, the $\mathcal{O}_P$-module Ω_P is generated by elements of $H^0(V, \Omega)$. Then every regular map f from the projective line Λ to V is constant.*

PROOF. Let $\omega \in H^0(V, \Omega)$; the differential form $f^*(\omega)$ is everywhere regular on Λ, thus zero. By virtue of the hypothesis on V, this implies that the tangent map to f is everywhere zero. In characteristic 0, this implies that f is constant. In characteristic $p > 0$, this implies that f factors as

$$\Lambda \xrightarrow{F} \Lambda \xrightarrow{g} V, \qquad \text{where } \Lambda \xrightarrow{F} \Lambda \text{ is the map } \lambda \to \lambda^p.$$

Applying the same argument to g, we deduce the existence of factorizations of f of the form $f = h \circ F^n$ with n an arbitrary integer, which is only possible when f is a constant map (indeed, otherwise the image C of Λ by f would be a curve, and we would necessarily have $p^n \leq [k(V) : k(C)]$). ☐

Corollary. *Every rational map from Λ to an Abelian variety is constant.*

PROOF. Indeed, since an Abelian variety is *complete*, such a map is everywhere regular, and on the other hand it is evident that an Abelian variety (and more generally any algebraic group, cf. no. 11) satisfies the condition of the lemma. ☐

(Observe that the same lemma furnishes a proof of Lüroth's theorem: it suffices to take for V a curve of genus > 0.)

We now pass to the proof of case a) of theorem 1:

Proposition 13. *Suppose that G is an extension of an Abelian variety A by the multiplicative group $\mathbf{G}_m$. If $\mathfrak{m} = \sum_{P \in S} P$, then $\mathfrak{m}$ is a modulus for f.*

PROOF. We are going to apply the criterion of proposition 9. Thus let g be a non-constant rational function on X such that

$$g \equiv 0 \bmod \mathfrak{m}, \qquad \text{i.e., } g(P) = 0 \text{ for } P \in S.$$

The map $\mathrm{Tr}_g\, f$ is a regular map from $\Lambda - \{0\}$ to G; composing it with the projection $G \to A$, we get a rational map from Λ to A which is constant according to the corollary to lemma 7. Thus $\mathrm{Tr}_g\, f$ takes its values in a class modulo $\mathbf{G}_m$, a class which is biregularly isomorphic to $\mathbf{G}_m$. Then applying lemma 6 we deduce that $\mathrm{Tr}_g\, f$ is constant, as was to be proved. ☐

Remark. When $G = \mathbf{G}_m$ the proof above gives the first half of prop. 6 (the determination of the modulus $\mathfrak{m}$), but not the second (the explicit value of the local symbol).

9. Proof in characteristic $p > 0$: reduction of case b) to the unipotent case

Suppose that we are in case b), that is to say that G contains a unipotent subgroup U such that $G/U = A$ is an Abelian variety. Writing U additively, there exists an integer r such that $p^r u = 0$ for all $u \in U$. This can be seen either by using the fact that U is a multiple extension of groups $\mathbf{G}_a$, or by writing u in the form $1 + n$, where n is a nilpotent matrix in $\mathbf{GL}_m$: then

$$(1 + n)^{p^r} = 1 + n^{p^r} = 1 \qquad \text{when } p^r \geq m.$$

On the other hand, according to a theorem of Weil ([89], p. 127), $p^r A = A$.

So put $U' = G/p^r G$, $G' = A \times U'$, and let $\theta : G \to G'$ be the product of the canonical maps $G \to A$ and $G \to U'$. The group U' is a *unipotent group*. Indeed, it satisfies $p^r U' = 0$, thus admits no subgroup isomorphic to $\mathbf{G}_m$, nor any subgroup isomorphic to a non-zero Abelian variety. Our assertion then follows from the structure theorems of no. 7. Furthermore, *the kernel of θ is finite*. Indeed, this kernel is equal to $U \bigcap p^r G$, and $p^r G$ is an Abelian variety (since $p^r U = 0$ the map $p^r : G \to G$ factors as $G \to A \to G$), thus $U \bigcap p^r G$, being the intersection of an affine variety with a complete variety, is indeed a finite set.

We have already checked theorem 1 in the case of an Abelian variety (this is essentially the content of the corollary to lemma 7); suppose it is true for a unipotent group. Then it will also hold for G', thus also for G by virtue of the following proposition:

Proposition 14. *Let θ be a homomorphism* (in the sense of algebraic groups) *from a commutative algebraic group G to a commutative algebraic group G' and suppose that the kernel of θ is finite. Then if $\mathfrak{m}$ is a modulus for $\theta \circ f$, it is also a modulus for f.*

(This is a converse to prop. 2 of no. 2.)

PROOF. We apply the criterion of proposition 9. If g is a non-constant rational function on X such that $g \equiv 0 \bmod \mathfrak{m}$, the map

$$\mathrm{Tr}_g(\theta \circ f) : \Lambda - \{0\} \to G'$$

is a constant map. But this map can also be written $\theta \circ \mathrm{Tr}_g f$, which shows that $\mathrm{Tr}_g f$ takes it values in a class modulo the kernel of θ, i.e., in a *finite* set. As $V - \{0\}$ is connected it follows that $\mathrm{Tr}_g f$ is constant, which proves the proposition. $\qquad\qquad\square$

Thus everything comes down to proving theorem 1 *when G is a unipotent group*. This is what will be done in the next no.

10. End of the proof: case where G is a unipotent group

We now suppose that G is a *connected, commutative, unipotent group*; we can thus consider it as embedded in a linear group $\mathbf{GL}_r(k)$, such that each of its elements is of the form $1 + N$, where $N = (n_{ij})$ is a matrix satisfying $n_{ij} = 0$ if $i \geq j$.

In particular, we can consider G as embedded in the vector space of all matrices of degree r over k, call it $\mathbf{M}_r(k)$. Every rational map g from a curve Y to the group G thus determines (and is determined by) r^2 rational functions g_{ij}, $i, j = 1 \ldots, r$. If Q is a point of Y, we put

$$w_Q(g) = \text{Sup}(0, -v_Q(g_{ij})), \qquad i, j = 1, \ldots r. \tag{1}$$

This applies in particular to $f : X - S \to G$. Then choose an integer $n > 0$ such that

$$n > (r-1)w_P(f) \qquad \text{for all } P \in S. \tag{2}$$

Proposition 15. *The modulus* $\mathfrak{m} = \sum_{P \in S} nP$ *is a modulus for* f.

PROOF. We are again going to apply the criterion of proposition 9. Thus let g be a non-constant rational function on X such that $v_P(g) \geq n$ for all $P \in S$. We must show that $f' = \text{Tr}_g\, f$ is a constant map; it will suffice for this to see that

$$w_0(f') < 1. \tag{3}$$

Indeed, this inequality will show that all of the $v_0(f'_{ij})$ are ≥ 0, in other words that f' is regular at the point 0 of Λ. The map f', being a regular map from the complete variety Λ to the affine variety G, will indeed be constant.

Let Y be a normal covering of Λ dominating X (cf. no. 13) and let π be the projection $Y \to X$. For each point $P \in X$ (resp. $Q \in Y$), let e_P (resp. e_Q) be the ramification index of P (resp. Q) in the covering $g : X \to \Lambda$ (resp. in the covering $\pi : Y \to X$). Choose a point $Q \in Y$ mapping to $P \in S$ and such that $e = e_Q$ is maximal for all the points Q having this property. Put $P = \pi(Q)$, $P \in S$. Then

$$w_Q(f' \circ g \circ \pi) = e e_P w_0(f')$$

whence

$$w_Q(f' \circ g \circ \pi) \geq ne\, w_0(f') \tag{4}$$

since $e_P = v_P(g)$ is $\geq n$.

Thus everything comes down to evaluating $w_Q(f' \circ g \circ \pi)$. But, if p^k denotes the inseparable degree of the extension $k(X)/k(\Lambda)$,

$$f' \circ g \circ \pi = p^k \sum f \circ \pi \circ \sigma_j \tag{5}$$

the σ_j denoting certain elements of the Galois group of Y (cf. no. 13). We are going to use the following lemma:

Lemma 8. *If* f^α *are rational maps from a curve* Y *to* G, *then*

$$w_Q\left(\sum f^\alpha\right) \leq (r-1)\,\text{Sup}\,w_Q(f^\alpha).$$

We admit for a moment this lemma and apply it to (5). We find

$$
\begin{aligned}
w_Q(f' \circ g \circ \pi) &\leq (r-1)\,\text{Sup}_j\, w_Q(f \circ \pi \circ \sigma_j) \\
&\leq (r-1)\,\text{Sup}_j\, w_{Q_j}(f \circ \pi) \quad \text{with } Q_j = \sigma_j(Q), \tag{6} \\
&\leq (r-1)\,\text{Sup}_j\, e_{Q_j} w_{P_j}(f) \quad \text{with } P_j = \pi(Q_j).
\end{aligned}
$$

If $P_j \notin S$, we have $w_{P_j}(f) = 0$, and we can thus consider in the right hand side of (6) only the terms corresponding to $P_j \in S$. For these, we have $e_{Q_j} \leq e$ and $w_{P_j}(f) < n/(r-1)$ in view of the choice of n. We thus deduce from (6) the inequality

$$w_Q(f' \circ g \circ \pi) < en \tag{7}$$

which, together with (4), proves (3).

It remains to prove lemma 8:

For that, we go back to writing the composition law on G *multiplicatively*. We must consider the product

$$\prod f^\alpha = \prod (1 + n^\alpha), \tag{8}$$

n^α being a rational map of Y into the space of matrices N such that $n_{ij} = 0$ for $i \geq j$. But a product of r such matrices is equal to zero. We can thus expand the product (8) in the form

$$\sum_{0 \leq k \leq r-1} \sum_{\alpha_1 < \cdots < \alpha_k} n^{\alpha_1} \cdots n^{\alpha_k}. \tag{9}$$

Formula (9) shows that the components $(\prod f^\alpha)_{ij}$ of $\prod f^\alpha$ are polynomials in those of n^α of total degree $\leq r-1$. Taking into account the definition of w_Q this gives:

$$w_Q\left(\prod f^\alpha\right) \leq (r-1) \operatorname{Sup} w_Q(n^\alpha) = (r-1) \operatorname{Sup} w_Q(f^\alpha), \tag{10}$$

as was to be shown.

This finishes the proof of proposition 15, and with it, the proof of theorem 1. □

Remark. One can also reduce the case of unipotent groups to Witt groups (no. 3), by using the fact that the group G is isogeneous to a product of Witt groups (cf. chap. VII, §2).

§3. Auxiliary results

11. Invariant differential forms on an algebraic group

If X is a non-singular algebraic variety, we write $\mathbf{T}_X$ for the fibre space of tangent vectors on X and $\mathbf{T}_X^*$ for the dual fibre space. Recall (cf. chap. II, no. 7) that $\mathbf{T}_X$ is defined by the condition that the sheaf $\mathcal{S}(\mathbf{T}_X^*)$ of germs of sections of $\mathbf{T}_X^*$ be isomorphic to the sheaf $\underline{\Omega}_X$ of differentials $D_k(\mathcal{O}_P)$ of the local rings of the points of X.

If $f : X \to Y$ is a regular map from the variety X to the variety Y (X and Y being supposed non-singular), f defines a homomorphism $\mathcal{O}_Y \to \mathcal{O}_X$, thus, passing to differentials, a homomorphism

$$\underline{\Omega}_Y \to \underline{\Omega}_X.$$

But, in general, if E_X and E_Y are vector bundles over X and Y respectively, every $\mathcal{O}_Y$-linear homomorphism from $\mathcal{S}(E_Y)$ to $\mathcal{S}(E_X)$ corresponds to a homomorphism of the fibre space E_X^* to E_Y^* compatible with f, and conversely; this can be immediately checked by using local coordinates. Applying this to $E_X = T_X^*$ and $E_Y = T_Y^*$ we see that f^* defines (and is defined by) a homomorphism

$$df : T_X \to T_Y$$

called the tangent map to f. Thus, by definition

$$\langle df(t), \omega \rangle = \langle t, f^*(\omega) \rangle,$$

if $t \in \mathbf{T}_X(x)$ and $\omega \in \mathbf{T}_Y^*(y)$, with $y = f(x)$ (we write $E(x)$ for the fibre over the point x of the fibre space E).

In other words, $\mathbf{T}_X$ is a *covariant functor* in X. It also enjoys the following property: if X and Y are two non-singular varieties, the canonical map $\mathbf{T}_{X \times Y} \to \mathbf{T}_X \times \mathbf{T}_Y$ is an isomorphism (indeed, the existence of the injections $X \to X \times Y$ and $Y \to X \times Y$ shows that this map is surjective, and as the two fibre spaces have the same dimension, it is an isomorphism).

These elementary properties will suffice for us. First consider a map $h : X \times Y \to Z$, and, for every $y \in Y$, write h_y for the partial map $x \to h(x, y)$. We have:

Lemma 9. *Suppose that X, Y and Z are non-singular. The map*

$$(t, y) \to dh_y(t), \qquad t \in \mathbf{T}_X, \ y \in Y,$$

is then a regular map from $\mathbf{T}_X \times Y$ to $\mathbf{T}_Z$.

PROOF. Indeed, this map factors as

$$\mathbf{T}_X \times Y \to \mathbf{T}_X \times \mathbf{T}_Y = \mathbf{T}_{X \times Y} \to \mathbf{T}_Z,$$

where $\mathbf{T}_{X \times Y} \to \mathbf{T}_Z$ is dh, while $\mathbf{T}_X \times Y \to \mathbf{T}_X \times \mathbf{T}_Y$ is the product of the identity map of $\mathbf{T}_X$ and the map from Y to $\mathbf{T}_Y$ which maps y to 0_y, the identity element of $\mathbf{T}_Y(y)$.   □

Now let G be an algebraic group; if $g \in G$, write ρ_g for the left translation $x \to g.x$. It is an automorphism of G (as an algebraic variety). It follows that $d\rho_g : \mathbf{T}_G(x) \to \mathbf{T}_G(g.x)$ is an isomorphism for every $x \in G$. If $t \in \mathbf{T}_G(x)$, we will write $g.t$ instead of $d\rho_g(t)$. According to the preceding lemma, $g.t$ is a regular function of g and t. In particular, let $\{t_1, \ldots, t_n\}$ be a basis of $\mathbf{T}_G(e)$ where e is the identity element of G. For each i, the map $g \to g.t_i$ is thus a regular section of the fibre space $\mathbf{T}_G$, and the $g.t_i$ form a basis of $\mathbf{T}_G(g)$ for all $g \in G$. In other words:

Proposition 16. *The space* $\mathbf{T}_G$ *of tangent vectors to the n-dimensional algebraic group* G *is a trivial fibre space, admitting for a frame n vector fields* $g.t_i$ *which are invariant by left translation.*

Passing to the dual fibre space $\mathbf{T}_G^*$ we get:

Corollary 1. *There exists a frame of* T_G^* *formed by n left-invariant differential forms* ω_i.

Now let $f : G \to G'$ be a homomorphism from the group G to the group G'; if $s(G)$ (resp. $s(G')$) denotes the k-vector space of left-invariant differential forms on G (resp. on G'), then $f^*(\omega) \in s(G)$ for every form $\omega \in s(G')$. With these notations:

Corollary 2. *Suppose that* $f : G \to G'$ *is surjective. In order that the algebraic group structure of* G' *be the quotient of that of* G, *it is necessary and sufficient that* $f^* : s(G') \to s(G)$ *be injective.*

PROOF. After replacing G by a quotient, we can suppose that f is bijective; we can also suppose that G and G' are connected. Let K/K' be the field extension corresponding to f. Since f is bijective, it is a purely inseparable extension; thus f is an isomorphism if and only if K/K' is separable. But one knows (see the criterion of separability of [11], exposé 13) that f is separable if and only if the tangent map to f is surjective on a non-empty open of G; in view of cor. 1, this condition is equivalent to $f^* : s(G') \to s(G)$ being injective, as was to be shown. $\square$

Corollary 3. *Suppose that* $f : G \to G'$ *is injective. In order that the algebraic group structure of* G *be induced by that of* G' *it is necessary and sufficient that* $f^* : s(G') \to s(G)$ *be surjective.*

PROOF. After replacing G' by a subgroup, we can suppose that f is bijective, and we are then reduced to the preceding corollary. $\square$

One could go further and define the adjoint linear representation of G, bi-invariant differential forms, the p-th power operation on the Lie algebra (in characteristic p), etc. None of this presents any difficulties. We limit ourselves to making explicit a property which was used in the proof of proposition 10:

Proposition 17. *Let* f *and* g *be two regular maps of an algebraic variety* X *to a commutative algebraic group* G. *If* ω *denotes an invariant differential form on* G, *then*

$$(f + g)^*(\omega) = f^*(\omega) + g^*(\omega).$$

PROOF. Denote by pr_1 and pr_2 the two projections of the group $G \times G$ to G and put $\rho = pr_1 + pr_2$. Since G is Abelian these maps are homomorphisms and the differentials $\rho^*(\omega)$, $pr_1^*(\omega)$ and $pr_2^*(\omega)$ are invariant differentials on $G \times G$. Since at the identity element of $G \times G$ we evidently have

$$\rho^*(\omega) = pr_1^*(\omega) + pr_2^*(\omega),$$

this equality is true everywhere. Let $(f, g) : X \to G \times G$ be the map defined by the pair (f, g). Then

$$f = pr_1 \circ (f, g), \qquad g = pr_2 \circ (f, g), \qquad f + g = \rho \circ (f, g),$$

whence

$$\begin{aligned}(f + g)^*(\omega) &= (f, g)^* \rho^*(\omega) \\ &= (f, g)^* pr_1^*(\omega) + (f, g)^* pr_2^*(\omega) \\ &= f^*(\omega) + g^*(\omega). \qquad \square\end{aligned}$$

12. Quotient of a variety by a finite group of automorphisms

Let V be an algebraic variety and let R be an equivalence relation on V. Denote by V/R the quotient set of V by R and let $\theta : V \to V/R$ be the canonical projection from V to V/R. We give V/R the quotient topology, where V has the Zariski topology. If f is a function defined in a neighborhood of $w \in V/R$, we say that f is regular at w if $f \circ \theta$ is regular in a neighborhood of $\theta^{-1}(w)$. The functions regular at w form a ring $\mathcal{O}_{V/R,w}$ and thus we get a subsheaf $\mathcal{O}_{V/R}$ of the sheaf of germs of functions on V/R. If the topology and the sheaf above satisfy the axioms for an algebraic variety [axioms (VA_I) and (VA_{II}) of FAC, no. 34] we say that the algebraic variety V/R is the *quotient variety* of V by R. The regular maps of V/R to an algebraic variety V' are then identified with the regular maps from V to V' which are constant on the classes of R. The structure of algebraic variety on V/R is thus the *quotient* of that of V, in the sense of Bourbaki, *Ens.* IV, §2, no. 6.

All this applies in particular to the case where R is the equivalence relation defined by a finite group $\mathfrak{g}$ acting on V. In this case, we write $V/\mathfrak{g}$ instead of V/R. We propose to show that, using a very broad hypothesis, $V/\mathfrak{g}$ is in fact an algebraic variety. We first will study a particular case:

Proposition 18. *Suppose that V is an affine variety with coordinate ring A. Then $V/\mathfrak{g}$ is an affine algebraic variety whose coordinate ring is naturally identified with the subring $A^{\mathfrak{g}}$ of elements of A fixed by $\mathfrak{g}$.*

PROOF. The ring A is a k-algebra which is generated by a finite number of elements $x_1, \ldots, x_n$. These elements are *integral* over $A^{\mathfrak{g}}$, as the equation of integral dependence

$$\prod_{\sigma \in \mathfrak{g}} (x - x^{\sigma}) = 0$$

shows. The following lemma then shows that $A^{\mathfrak{g}}$ is a k-algebra of finite type:

Lemma 10. *Let A be an algebra of finite type over a commutative Noetherian ring k and let B be a subalgebra of A such that every element of A is integral over B. Then B is a k-algebra of finite type.*

PROOF. Let x_i, $1 \leq i \leq n$, be generators of the algebra A; each of these elements satisfies an equation of integral dependence over B, say $f_i(x_i) = 0$. Let $b_1, \ldots, b_r$ be the coefficients of the these equations and let $C = k[b_1, \ldots, b_r]$ be the subalgebra of B generated by the b_j. The x_i are integral over C and generate A; this implies that A is a C-module of finite type. But C is Noetherian. Thus B, which is a C-submodule of A, can be generated by a finite number of elements $y_1, \ldots, y_m$, and we have $B = k[b_1, \ldots, b_r, y_1, \ldots, y_m]$, as was to be proved. $\square$

We now return to the proof of proposition 18.

As B has no nilpotent elements, there exists an affine variety W whose coordinate ring is B. The inclusion $B \to A$ defines a regular map $\theta : V \to W$ which is surjective (because A is integral over B) and invariant by $\mathfrak{g}$. Further, if v and v' are points of V which are not equivalent modulo $\mathfrak{g}$, one can find $f \in A$ such that

$$f(v) = 0 \quad \text{and} \quad f(v'^{\sigma}) = 1$$

for every $\sigma \in \mathfrak{g}$. The function $F = \prod f^{\sigma}$ then belongs to $A^{\mathfrak{g}}$ and satisfies $F(v) = 0$, $F(v') = 1$, which shows that $\theta(v) \neq \theta(v')$. Thus θ identifies the quotient set $V/\mathfrak{g}$ with W. We check that the structure of algebraic variety on W is in fact the quotient of that of V. First of all, the topology of W is the quotient topology of V: indeed, it suffices to see that, if V' is closed in V, $\theta(V')$ is closed in W. But, if V' is defined by the vanishing of functions $f_i \in A$, it is clear that $\theta(V')$ is defined by the vanishing of $F_i = \prod_{\sigma \in \mathfrak{g}} f_i^{\sigma}$, which indeed belong to $A^{\mathfrak{g}}$. Finally, if g is a function invariant by $\mathfrak{g}$ and regular on a saturated neighborhood of $v \in V$, g can be written in the form a/s, with $a \in A$, and $s \neq 0$ on the orbit of v. After replacing s by $S = \prod s^{\sigma}$, we can assume that $s \in A^{\mathfrak{g}}$, whence $a \in A^{\mathfrak{g}}$ (at least if V is irreducible—otherwise, the argument must be modified in the obvious way). This shows that g is of the form $g' \circ \theta$, with g' regular in a neighborhood $\theta(v)$; the local rings of W thus coincide with those of $V/\mathfrak{g}$, as was to be shown. $\square$

Corollary.
a) *If $\mathfrak{g}$ and $\mathfrak{g}'$ act on affine varieties V and V', then*

$$(V \times V')/(\mathfrak{g} \times \mathfrak{g}') = V/\mathfrak{g} \times V'/\mathfrak{g}'.$$

b) *If V is irreducible, the same is true of $V/\mathfrak{g}$, and*

$$k(V/\mathfrak{g}) = k(V)^\mathfrak{g}.$$

c) *If V is normal, the same is true of $V/\mathfrak{g}$.*

PROOF. Assertion a) is equivalent to the formula

$$(A \otimes_k A')^{\mathfrak{g} \times \mathfrak{g}'} = A^\mathfrak{g} \otimes_k A'^{\mathfrak{g}'}$$

which is in fact valid for any vector spaces A and A'.

Assertions b) and c) follow from the classical formula

$$A^\mathfrak{g} = A \bigcap k(V)^\mathfrak{g}.$$

We leave the details of these verifications to the reader. □

Proposition 19. *Let $\mathfrak{g}$ be a finite group acting on an algebraic variety V and suppose that every orbit of $\mathfrak{g}$ is contained in an affine open of V. Then $V/\mathfrak{g}$ is an algebraic variety.*

PROOF. If S is an orbit of $\mathfrak{g}$, there exists an affine open U containing S; putting $U' = \bigcap U^\sigma$, the open U' is an affine open which is invariant by $\mathfrak{g}$ and still contains S. Cover V by such opens U'_i, and let $U_i = U'_i/\mathfrak{g}$ be their images in $V/\mathfrak{g}$; the U_i are affine opens, by virtue of prop. 18. The axiom (VA_I) of algebraic varieties is thus satisfied. It remains to check (VA'_{II}), in other words that $\Delta_{V/R} \bigcap (U_i \times U_j)$ is closed in $U_i \times U_j$ ($\Delta_{V/R}$ denoting the diagonal of V/R). But this set is the image of $\Delta_V \bigcap (U'_i \times U'_j)$ by the canonical projection

$$U'_i \times U'_j \to U_i \times U_j = (U'_i \times U'_j)/(\mathfrak{g} \times \mathfrak{g})$$

(cf. the preceding corollary); as V is an algebraic variety, $\Delta_V \bigcap (U'_i \times U'_j)$ is closed and the same is true of the preceding map, which finishes the proof. □

Examples. 1) The condition of prop. 19 is always satisfied if V is a locally closed subvariety of a projective space $\mathbf{P}_r(k)$. Indeed, let S be an orbit of $\mathfrak{g}$ and let $\overline{V}$ be the closure of V in $\mathbf{P}_r(k)$; the set $F = \overline{V} - V$ is closed and does not meet S. We can thus find a homogeneous form $\varphi(x_0, \ldots, x_r)$ vanishing on F and not zero on the points of S. The set U of points of V where $\varphi(x_0, \ldots, x_r) \neq 0$ is then an open of V satisfying the condition (cf. FAC, no. 52).

2) The same is true when V is a *group variety* G. Indeed, suppose at first that G is connected and let g_i be the points of the orbit S considered. If U_0 is a non-empty affine open of G, the intersection of the $g_i^{-1} U_0$ is not empty; let g be an element of this intersection. One immediately checks that $U = U_0 g^{-1}$ answers the question. The case where G is not connected can be reduced to the preceding case.

(Note that in fact Barsotti [3] has proved that every algebraic group can be embedded in a projective space; the same result holds for homogeneous spaces according to Chow [19].)

3) However, Nagata [58] has constructed an algebraic variety V on which the group $G = \mathbf{Z}/2\mathbf{Z}$ acts where the quotient $V/\mathfrak{g}$ is not an algebraic variety.

Remarks. 1) In view of its local character, the corollary to proposition 18 remains valid for (not necessarily affine) varieties which satisfy the condition of proposition 19.

2) Let k' be a subfield of k and suppose that V is given the structure of k'-variety (embeddable in a projective space, to fix ideas). If the actions of $\mathfrak{g}$ are defined over k', then $V/\mathfrak{g}$ can be given, in a canonical way, a structure of k'-variety. This can be seen immediately by reducing to the affine case and using the evident formula

$$(A' \otimes_{k'} k)^{\mathfrak{g}} = A'^{\mathfrak{g}} \otimes_{k'} k$$

valid for every k'-vector space A'.

3) It is perhaps worth mentioning that one of the "intuitive" properties of quotients is not true: if V' is a subvariety of V stable by $\mathfrak{g}$, the quotient variety $V'/\mathfrak{g}$ is not necessarily identified with a subvariety of $V/\mathfrak{g}$ (the map $V'/\mathfrak{g} \to V/\mathfrak{g}$ is purely inseparable, as we will see an example of in no. 14). However, this is always true in characteristic 0, or if $\mathfrak{g}$ acts without fixed points.

13. Some formulas related to coverings

Although we will need these results only in the case of curves, we prove them for varieties of arbitrary dimension, which offers no extra difficulties.

Let X' be a *non-singular* irreducible variety, let $F' = k(X')$ be its field of rational functions, and let F/F' be a finite algebraic extension. Denote by X the *normalization* of X' in the extension F/F' (cf. for example [11], exposé 7, or [51], chap. V); it is an irreducible normal variety, endowed with a projection $g : X \to X'$; one has $k(X) = F$.

Let $P' \in X'$; the points $P \in X$ mapping to P' are finite in number. Since P' is *simple* on X', intersection theory applies and permits us to define a multiplicity e_P, whence a cycle $g^{-1}(P') = \sum_{P \to P'} e_P P$. We recall rapidly the definition of the integer e_P:

Let A and A' be the local rings of P and P'; then $A' \subset A$. The maximal ideal $\mathfrak{m}'$ of A' generates a primary ideal $\mathfrak{m}'A$ in A and

$$e_P = e_A(\mathfrak{m}'A) = \text{the multiplicity of the ideal } \mathfrak{m}'A \text{ in } A,$$

in the sense of Chevalley-Samuel [70].

One can also show that e_P is equal to the alternating sum of the dimensions of the k-vector spaces $\text{Tor}_i^{A'}(A, k)$; this is a particular case of the "Tor formula".

In the case of curves, A and A' are discrete valuation rings, and e_P is equal to the ramification index of the corresponding valuations.

Now let L be a normal extension of F' containing F, and let $\mathfrak{g}$ be the group of F'-automorphisms of L (cf. Bourbaki, *Alg.* V, §10, no. 9); let $\mathfrak{h}$ be the subgroup of $\mathfrak{g}$ formed by automorphisms fixing all the elements of F. If Y is the normalization of X' in L (or that of X, it is the same thing), there are projections

$$Y \xrightarrow{\pi} X \xrightarrow{g} X'.$$

The group $\mathfrak{g}$ is a group of automorphisms of Y.

Proposition 20. *For every $P \in X$, we have $g^{-1}(g(P)) = [F : F']_i \sum \pi \circ \sigma_i(Q)$, where $Q \in Y$ is such that $\pi(Q) = P$ and where the σ_i denote representatives in $\mathfrak{g}$ of the elements of $\mathfrak{g}/\mathfrak{h}$.*

(Recall that $[F : F']_i$ denotes the *inseparable part* of the degree of F/F'.)

PROOF. Let $P' = g(P)$ and let $h = g \circ \pi$. First we determine the cycle $h^{-1}(P')$. It is a linear combination $\sum_{\sigma \in \mathfrak{g}} n_\sigma \sigma(Q)$; as the σ are automorphisms of Y compatible with h, the n_σ are equal to the same integer n. The *projection formula* (cf. [70], p. 32) shows that $h(h^{-1}(P')) = [L : F']P'$, whence $\deg(h^{-1}(P')) = [L : F']$ which determines the integer n. Remarking that $[L : F']$ is equal to the product of $[L : F']_i$ by the order $[\mathfrak{g}]$ of $\mathfrak{g}$, we get

$$h^{-1}(P') = [L : F']_i \sum_{\sigma \in \mathfrak{g}} \sigma(Q).$$

If we now apply the projection formula to π, we get

$$\pi(h^{-1}(P')) = [L : F]g^{-1}(P'). \tag{*}$$

But the sum $\sum_{\sigma \in \mathfrak{g}} \sigma(Q)$ can be written $\sum_{\alpha \in \mathfrak{h}} \sum_i \alpha \circ \sigma_i(Q)$, and since $\pi \circ \alpha = \pi$ for all $\alpha \in \mathfrak{h}$, we have

$$\pi(h^{-1}(P')) = [L : F']_i.[\mathfrak{h}]. \sum \pi \circ \sigma_i(Q).$$

As $[L : F] = [L : F]_i.[\mathfrak{h}]$ and $[L : F']_i = [L : F]_i.[F : F']_i$, we finally get

$$\pi(h^{-1}(P')) = [L : F].[F : F']_i. \sum \pi \circ \sigma_i(Q) \tag{**}$$

and the proposition follows by comparing (*) and (**).   □

Remark. Suppose that the covering $g : X \to X'$ is *separable*, i.e., that $[F : F']_i = 1$. If $f \in k(X)$, one then has the following formula, analogous to that of prop. 20:

$$(\text{Tr}_g f) \circ \pi = \sum f \circ \pi \circ \sigma_i.$$

The same formula is valid for differential forms (for one writes them $\sum f_i \omega'_i$ with $f_i \in k(X)$ and ω'_i a differential form on X' and applies the formula to the functions f_i).

14. Symmetric products

Let X be an algebraic variety; in this no. and the following we suppose that *every finite subset of X is contained in an affine open* (cf. no. 12). Then let n be an integer ≥ 1 and denote by $\mathcal{S}_n$ the symmetric group of degree n; letting $\mathcal{S}_n$ act on X^n by permuting the factors, the pair $(X^n, \mathcal{S}_n)$ satisfies the conditions of proposition 19. The quotient $X^n/\mathcal{S}_n$ is thus an algebraic variety called *the n-fold symmetric product of X*, which we denote by $X^{(n)}$.

By the very definition, a point $M \in X^{(n)}$ can be identified with a formal sum $\sum_{P \in X} n_P P$ with $n_P \geq 0$ and $\sum n_P = n$, in other words with a *positive cycle of dimension 0 and degree n*. When X is a projective variety, one easily checks that $X^{(n)}$ is isomorphic to the variety of "Chow points" of the cycles in question.

If X is irreducible, the same is evidently true of $X^{(n)}$; similarly for X normal. If X is a *non-singular curve*, $X^{(n)}$ is non-singular. This can be seen either by determining explicitly the completions of the local rings of $X^{(n)}$, or by first treating the case where X is a projective line (in this case $X^{(n)} = \mathbf{P}_n(k)$, which is non-singular) and using the fact that every curve is a covering of a line and that one can require this covering to be non-ramified at given points. However, if X is a variety of dimension ≥ 2 and if $n \geq 2$, the symmetric product $X^{(n)}$ always has singularities.

Sending every point $P \in X$ to the point of $X^{(n)}$ corresponding to the cycle nP, we get an injective map $\delta_n : X \to X^{(n)}$. This map is regular, since it is the composition of the diagonal map $X \to X^n$ with the canonical projection $X^n \to X^{(n)}$. In view of the definition of the topology of $X^{(n)}$, it is even a homeomorphism from X to a subvariety of $X^{(n)}$. When n is prime to the characteristic p of the base field, one easily sees that this homeomorphism is biregular. This is not true in general, as the following proposition shows:

Proposition 21. *If $n = p^m$ with $m \geq 0$, the sheaf of functions induced on $\delta_n(X)$ by the regular functions of $X^{(n)}$ coincides with the sheaf of p^m-th powers of $\mathcal{O}_X$ transported by δ_n.*

(Let X_m be the algebraic variety obtained by giving X the sheaf of p^m-th powers of the regular functions of X; the proposition means that δ_n defines a biregular isomorphism from X_m to $\delta_n(X)$, with the structure induced by that of $X^{(n)}$.)

PROOF. The question being local, we can assume that X is affine. Let A be the coordinate ring of X; that of X^n is the n-th tensor power of A, call it A_n, and that of $X^{(n)}$ is the subring of A_n formed by elements invariant by the group $\mathcal{S}_n$. We are thus reduced to proving the following result:

Lemma 11. *Let A be a commutative algebra over a field k of characteristic p and let $n = p^m$ be any power of p. Let $A_n = A \otimes \cdots \otimes A$ (n times) and let B_n be the subalgebra of A_n formed by elements invariant under the symmetric group $\mathcal{S}_n$. Let $u : A_n \to A$ be the homomorphism defined by the formula*

$$u(a_1 \otimes \cdots \otimes a_n) = a_1 \ldots a_n.$$

Then $u(B_n) = k.A^n$ and $u(B_n) = A^n$ if k is perfect.

(By abuse of notation, we write A^n for the set of elements of A which are n-th powers; since n is a power of the characteristic, this is a subring of A.)

PROOF. As $u(a \otimes \cdots \otimes a) = a^n$, we have $u(B_n) \supset A^n$, and since u is a homomorphism of algebras, this implies that $u(B_n) \supset k.A^n$. We show conversely that $u(B_n) \subset k.A^n$. Let a_i be a basis of A. The products $\alpha = a_{i_1} \otimes \cdots \otimes a_{i_n}$ form a basis of A_n. Let t_α be the subgroup of $\mathcal{S}_n$ formed by permutations leaving α fixed and put (as usual in the theory of symmetric functions)

$$S(\alpha) = \sum \sigma_i(\alpha)$$

the σ_i running through a systems of representatives of $\mathcal{S}_n/t_\alpha$. It is clear that the $S(\alpha)$ form a basis of B_n, and it will thus suffice to prove that $u(S(\alpha)) \in A^n$ for all α. But

$$u(S(\alpha)) = n_\alpha a_{i_1} \ldots a_{i_n} \qquad \text{with} \qquad n_\alpha = (\mathcal{S}_n : t_\alpha).$$

So we distinguish two cases:

i) $a_{i_1} = \cdots a_{i_n} = a$, whence $t_\alpha = \mathcal{S}_n$ and $u(S(\alpha)) = a^n$.

ii) The a_{i_j} are not all equal, in which case t_α is a product of symmetric groups of degrees $< n = p^m$. The index of t_α in $\mathcal{S}_n$ is thus divisible by p. Indeed, otherwise a p-Sylow subgroup of t_α would also be a p-Sylow subgroup of $\mathcal{S}_n$. But $\mathcal{S}_n$ contains a cyclic subgroup of order p^m, while t_n evidently does not contain it; thus there is a contradiction. Since p divides $(\mathcal{S}_n : t_\alpha)$, $u(S(\alpha)) = 0$, which finishes the proof.   □

15. Symmetric products and coverings

Let $g : X \to X'$ be a covering satisfying the hypotheses of no. 13; the variety X' is thus irreducible and non-singular, and the variety X is its normalization in a finite extension F/F' of its function field. For every $P' \in X'$, the cycle $g^{-1}(P')$ is defined; its degree r is equal to $[F : F']$ and so one can consider it as an element of the symmetric product $X^{(r)}$, which we will denote by $g^*(P')$.

Proposition 22. *The map* $g^* : X' \to X^{(r)}$ *is everywhere regular.*

PROOF. Put $n = [F : F']_i$ and $s = [F : F']_s$ so that $ns = r$. The integer n is equal to a power p^m of the characteristic. We are going to begin by constructing a certain normal extension L of F' containing F, which will permit us to apply proposition 20.

Let F'' be the largest separable extension of F' contained in F and let G be a Galois extension of F' containing F'' (see the diagram below).

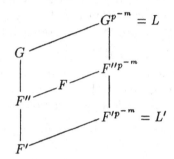

We take $L = G^{p^{-m}}$. The map $x \to x^{p^{-m}}$ shows that L is a Galois extension of $L' = F'^{p^{-m}}$ having the same Galois group $\mathfrak{g}$ as the extension G/F'. It is thus a normal extension of F'. Further, since

$$[F : F']_i = [F : F''] = p^m,$$

we have $F \subset F''^{p^{-m}} \subset L$.

Let Y (resp. Y') be the normalization of X' in the extension L/F' (resp. L'/F'). The projection $h : Y \to X'$ factors as $Y \xrightarrow{\pi} X \xrightarrow{g} X'$ and also as $Y \to Y' \to X'$. By construction, $Y' = Y/\mathfrak{g}$ and $X' = Y'_m$ (we write Y'_m for the variety obtained by giving Y' the sheaf of p^m-th powers of its local rings, cf. no. 14).

Now we apply proposition 20: if $\mathfrak{h}$ denotes the subgroup of $\mathfrak{g}$ associated to the field F'' and if the σ_i are representatives in $\mathfrak{g}$ of the elements of $\mathfrak{g}/\mathfrak{h}$, then

$$g^{-1}(P') = p^m \sum_{i=1}^{i=s} \pi \circ \sigma_i(Q), \qquad \text{with } h(Q) = P'.$$

We leave aside the factor p^m and consider the map $Q \to \sum \pi \circ \sigma_i(Q)$. It is a regular map from Y to the symmetric product $X^{(s)}$: indeed, we can factor it by regular maps $Y \to Y^s \to X^s \to X^{(s)}$. Furthermore, this map is invariant by $\mathfrak{g}$: it thus defines by passage to the quotient a regular map $\alpha : Y/\mathfrak{g} = Y' \to X^{(s)}$. Put $Z = X^{(s)}$ and let Z_m be the variety obtained by giving Z the sheaf of p^m-th powers of its local rings. The map α defines a regular map $\alpha_m : Y'_m = X' \to Z_m$. On the other hand, proposition 21 applied to Z furnishes a regular map $\delta'_n : Z_m \to Z^{(n)}$. The variety $Z^{(n)} = (X^{(s)})^{(n)}$ is the quotient of $X^{ns} = X^r$ by a subgroup of $\mathcal{S}_r$; thus there is a canonical projection $(X^{(s)})^{(n)} \xrightarrow{\gamma} X^{(r)}$. Forming the composition

$$ X' \xrightarrow{\alpha_m} Z_m \xrightarrow{\delta'_n} (X^{(s)})^{(n)} \xrightarrow{\gamma} X^{(r)}, $$

we get a regular map, and proposition 20 means that this map coincides with g^*.   □

Remarks. 1) The hypothesis that X is non-singular was only used so that $g^{-1}(P')$ makes sense. Indeed, if $g : X \to X'$ is any covering (X and X' being normal varieties), one can *define* $g^{-1}(P')$, for $P' \in X'$, by the formula of proposition 20, and the proof that we gave again applies.

2) In the case of curves, $F = F''^{p^{-m}}$, which slightly simplifies the proof. In any case, there are essentially two cases to consider: the separable case, which can be treated immediately by Galois arguments, and the purely inseparable case, which follows from proposition 21.

Bibliographic note

Theorem 1, which is the principal result of this chapter, is due to Rosenlicht [65]; we have reproduced his proof without essential change.

The local symbols of §1 are inspired by those which come up in class field theory (see in particular Schmid [72]) and the latter can be recovered from the former (cf. chap. VI, §6). The local symbol for the group G_a is due to Tate (non-published). The reciprocity formula $f((g)) = g((f))$ has been known for some time; it figures in the 1940 note of Weil [86] on the Riemann hypothesis and Igusa gave, in 1956, a direct proof [39]. Recently, Lang has shown that one can consider it as a particular case of a general reciprocity formula on Abelian varieties; see [52], chap. VI, §4.

"Chevalley's theorem" cited in no. 7 was proved by Barsotti [4] and Rosenlicht [66]. Chevalley himself never published his proof. The decomposition of a commutative connected linear group into a product of a torus and a unipotent group (prop. 12) is due to Kolchin [46]; see also Borel [9]. It would be interesting to avoid these structure theorems, as one can do in characteristic 0.

Tangent vectors and differential forms are treated in general in the literature from a "birational" point of view, which is insufficient. We limited ourselves to giving some elementary results that we needed; see also Rosenlicht [67].

The quotient varieties $V/\mathfrak{g}$ can be defined, in the normal case, using "Chow points." For the general case see [77] where there are also some other results on local rings.

The relations among symmetric products and coverings are nearly equivalent to the "symmetric functions theorem" of Weil ([89], §1). In both cases, the essential point is lemma 11.

CHAPTER IV

Singular Algebraic Curves

The principal goal of this chapter is to prepare for the construction of *generalized Jacobians* which is the object of the following chapter. Some of the results have their own interest because of their applications to the theory of surfaces.

Here again we do not treat questions of rationality, and we assume that the ground field k is algebraically closed.

§1. Structure of a singular curve

1. Normalization of an algebraic variety

We begin by rapidly reviewing the construction and elementary properties of the *normalization* of an algebraic variety. For more details, the reader is refered to the Cartan-Chevalley seminar [11] or to Lang [51].

Let X' be an irreducible algebraic variety, let $\mathcal{O}'$ be the sheaf of its local rings, and let $K = k(X')$ be the field of rational functions on X'. For every point $Q \in X'$ we denote by $\mathcal{O}_Q$ the *integral closure* of $\mathcal{O}'_Q$ in its field of fractions K. If U' is an affine open in X' and if $A' = H^0(U', \mathcal{O}')$ is the corresponding coordinate ring, the integral closure A of A' in K is an A'-module of finite type, corresponding to a normal affine variety U endowed with a canonical projection $p : U \to U'$. By glueing the varieties U we get a normal algebraic variety X which is called the *normalization* of X'; one immediately checks that, for every $Q \in X'$, $\mathcal{O}_Q = \bigcap_{P \to Q} \mathcal{O}_P$ where $\mathcal{O}_P$

denotes the local ring of a point $P \in X$ mapping to Q (these points are finite in number). The sheaf $\mathcal{O}$ on X' is thus the *direct image* $p_*(\mathcal{O}_X)$ of the sheaf $\mathcal{O}_X$ of local rings on X.

The annihilator $\mathfrak{c}$ of the sheaf of modules $\mathcal{O}/\mathcal{O}'$ is called the *conductor* of $\mathcal{O}$ into $\mathcal{O}'$. It is a coherent sheaf of ideals on X' and its variety S' is the set of points of X' which are not normal. Putting $S = p^{-1}(S')$, the projection p is a *biregular isomorphism* from $X - S$ to $X' - S'$. If Q is a point of X', the local ideal $\mathfrak{c}_Q$ is the set of $f \in \mathcal{O}'_Q$ such that $g \in \mathcal{O}_Q$ implies $fg \in \mathcal{O}'_Q$; one can also say that $\mathfrak{c}_Q$ is the largest ideal of $\mathcal{O}'_Q$ which is an ideal of $\mathcal{O}_Q$. There are inclusions

$$\mathcal{O}_Q \supset k + \mathfrak{r}_Q \supset \mathcal{O}'_Q \supset k + \mathfrak{c}_Q \tag{1}$$

where $\mathfrak{r}_Q$ denotes the *radical* of the semi-local ring $\mathcal{O}_Q$, in other words the set of $f \in \mathcal{O}_Q$ which take the value 0 at all the points P mapping to Q.

2. Case of an algebraic curve

If X and X' are *algebraic curves*, the sets S and S' have dimension 0, in other words they are *finite subsets* of X and X'. Furthermore, the set S' is nothing other than the set of *singular* points of X': indeed, one knows that, if $\mathfrak{o}$ is an integral local ring of dimension one, the conditions "$\mathfrak{o}$ is integrally closed" and "$\mathfrak{o}$ is regular" are equivalent.

The coherent sheaf $\mathcal{O}/\mathcal{O}'$ is concentrated on S'; it follows that, for every $Q \in X'$, $\dim(\mathcal{O}_Q/\mathcal{O}'_Q) < +\infty$. We put

$$\delta_Q = \dim(\mathcal{O}_Q/\mathcal{O}'_Q), \quad Q \in X'. \tag{2}$$

It follows from the preceding that:

Proposition 1. *The integer δ_Q is > 0 if and only if Q is a singular point of X'.*

Note that δ_Q is *invariant under completion*:

$$\delta_Q = \dim(\widehat{\mathcal{O}_Q}/\widehat{\mathcal{O}'_Q}). \tag{3}$$

This is immediate from the fact that $\mathcal{O}_Q/\mathcal{O}'_Q$ is finite dimensional.

It follows from (3) that two singular points which are "analytically isomorphic" (that is to say which have the same completed local ring) have the same δ_Q; in other words, δ_Q is an *analytic invariant*.

The fact that $\mathcal{O}_Q/\mathcal{O}'_Q$ is a finite dimensional k-vector space implies that the same is true of $\mathcal{O}'_Q/\mathfrak{c}_Q$, thus also of $\mathcal{O}_Q/\mathfrak{c}_Q$. Thus the ideal $\mathfrak{c}_Q$ contains a power $\mathfrak{r}_Q^n$ of the radical r_Q of $\mathcal{O}_Q$, and the inclusions (1) give inclusions

$$k + \mathfrak{r}_Q \supset \mathcal{O}'_Q \supset k + \mathfrak{r}_Q^n. \tag{4}$$

Remark. All of this is valid for an algebraic variety of any dimension, under the hypothesis that the set of its non-normal points is *finite*.

3. Construction of a singular curve from its normalization

In the preceding no., we started with X' to construct X; we are now going to go in the opposite direction.

Let X be an irreducible and non-singular algebraic curve, let $\mathcal{O}$ be the sheaf of its local rings, and let $K = k(X)$. We give ourselves a finite subset S of X and an equivalence relation R on S; if $S' = S/R$, we will define X' as the union of $X - S$ and S'. Thus there is a canonical projection $p : X \to X'$. If $Q \in X'$, we put $\mathcal{O}_Q = \bigcap_{P \to Q} \mathcal{O}_P$ and we denote by $\mathfrak{r}_Q$ the radical of $\mathcal{O}_Q$. We also give ourselves, for each $Q \in S'$, a subring $\mathcal{O}'_Q$ of $\mathcal{O}_Q$, distinct from $\mathcal{O}_Q$, satisfying the inclusions (4) for at least one integer n. If $Q \in X' - S'$, we put $\mathcal{O}'_Q = \mathcal{O}_Q$. The $\mathcal{O}'_Q$ form a subsheaf $\mathcal{O}'$ of the sheaf of functions on X' (we endow X' with the topology where the closed subsets distinct from X' are the finite subsets).

Proposition 2. *The sheaf $\mathcal{O}'$ endows X' with a structure of algebraic curve having X for its normalization and S' for its set of singular points.*

PROOF. We are going to show that, if X is an *affine* curve, the same is true of X'; the general case will follow immediately from that (cover X by affine opens).

Thus let A be the coordinate ring of X; for every $P \in X$, let $\mathfrak{a}_P$ be the maximal ideal of A formed by functions zero at P. Let $A' \subset K$ be the intersection of the $\mathcal{O}'_Q$ for Q running through X'; evidently $A' \subset A$. On the other hand, denoting by $\mathfrak{r}$ the intersection of the $\mathfrak{a}_P$ for $P \in S$, the conditions (4) show that A' contains $k + \mathfrak{r}^n$, for n large enough. As $\mathfrak{r}^n$ has finite codimension in A, the same is true of A'; thus A is an A' module of finite type, and A is integral over A'. It follows (chap. III, no. 12, lemma 10) that A' is a *k-algebra of finite type*, corresponding to an affine algebraic variety Y. As A is the integral closure of A' in K, there is a canonical projection $q : X \to Y$ which makes X the *normalization* of Y.

It remains to see that Y is isomorphic to X'. First of all, let P_1 and P_2 be two points of X having the same image Q in X'; these points correspond to homomorphisms $\varphi_i : A \to k$ $(i = 1, 2)$. Putting $\mathfrak{r}_Q = \bigcap_{P \to Q} \mathfrak{a}_P$, the homomorphisms φ_1 and φ_2 vanish on $\mathfrak{r}_Q$, thus coincide on $k + \mathfrak{r}_Q$ and a *fortiori* on A' according to (4). As the points of Y correspond bijectively to the homomorphisms $A' \to k$, it follows that P_1 and P_2 have the *same image* in Y. Conversely, suppose that P_1 and P_2 have distinct images Q_1 and Q_2 in X'. From the fact that the ideals $\mathfrak{a}_P$ are maximal ideals of A, one can, according to an elementary result, find $f \in A$ such that

$$\begin{cases} f \equiv 0 \bmod \mathfrak{a}_P^n & \text{if } P \to Q_1 \\ f \equiv 1 \bmod \mathfrak{a}_P^n & \text{if } P \to Q_2 \text{ or if } P \in S \text{ and } P \not\to Q_1. \end{cases} \tag{5}$$

The formulas (4) show that $f \in A'$, and as $f(P_1) = 0$ and $f(P_2) = 1$, the points P_1 and P_2 have distinct images in Y.

Thus the canonical map $X \to X'$ defines by passage to the quotient a *bijection* $X' \to Y$. We must now show that, identifying X' and Y by means of this bijection, the sheaves $\mathcal{O}'$ and $\mathcal{O}_Y$ coincide. The inclusion $\mathcal{O}_Y \subset \mathcal{O}'$ is evident. Thus let $f \in \mathcal{O}'_Q$, $Q \in X'$; we show that f can be written in the form

$$f = a/b \quad \text{with} \quad a \in A', \ b \in A', \ b(P) \neq 0 \text{ if } P \to Q. \tag{6}$$

Since $f \in \mathcal{O}_Q = \bigcap_{P \to Q} \mathcal{O}_P$ in any case

$$f = c/d \quad \text{with} \quad c \in A, \ d \in A, \ d(P) \neq 0 \text{ if } P \to Q. \tag{7}$$

One knows that there exists $t \in A$ such that

$$\begin{cases} t \equiv d^{-1} \mod \mathfrak{a}_P^n & \text{if} \quad P \to Q \\ t \equiv 0 \mod \mathfrak{a}_P^n & \text{if} \quad P \in S \text{ and } P \not\to Q. \end{cases}$$

The product td belongs to $\mathcal{O}'_R$ for every $R \in S'$, whence $td \in A'$, and $td(P) = 1$ for $P \to Q$; similarly, tc belongs to $\mathcal{O}'_R$ for all $R \in S'$ distinct from Q. On the other hand, $f = ct/dt$, and $f \in \mathcal{O}'_Q$ by hypothesis; we deduce that $ct \in \mathcal{O}'_Q$, and $ct \in A'$. If we put $a = ct$ and $b = dt$, the conditions (6) are indeed verified, which finishes the proof. $\qquad \square$

Remark. The situation studied above is analogous to what one encounters in arithmetic when A is the *ring of integers* of an algebraic number field K and A' is an *order* of K, that is to say a subring of A having K as its field of fractions. Here again, one has a *conductor* $\mathfrak{c}$ of A into A', the annihilator of the A'-module A/A'; the prime ideals of A' which enter into $\mathfrak{c}$ play the role of singular points.

4. Singular curve defined by a modulus

Let X be a complete, irreducible, non-singular curve and let $\mathfrak{m} = \sum n_P P$ be a "modulus" on X (cf. chap. III). We exclude the trivial case $\mathfrak{m} = 0$, as well as the case where $\mathfrak{m} = P$, and thus we suppose that $\deg(\mathfrak{m}) \geq 2$.

Let S be the *support* of $\mathfrak{m}$, in other words the set of $P \in X$ such that $n_P > 0$. We take for S' a set reduced to one point Q, and we put $X' = (X - S) \cup \{Q\}$ (X' is thus obtained from X by putting the points of S together into one). If $\mathfrak{c}_Q$ denotes the ideal of $\mathcal{O}_Q$ formed by the functions f such that $f \equiv 0 \mod \mathfrak{m}$, we put

$$\mathcal{O}'_Q = k + \mathfrak{c}_Q. \tag{9}$$

One then immediately checks that all the conditions of no. 3 are satisfied ($\mathcal{O}'_Q \neq \mathcal{O}_Q$ because of the hypothesis $\deg(\mathfrak{m}) \geq 2$). The singular curve associated to this data will be denoted $X_\mathfrak{m}$. It admits Q as its unique singular point and

$$\delta_Q = \dim \mathcal{O}_Q/(k + \mathfrak{c}_Q) = \dim \mathcal{O}_Q/\mathfrak{c}_q - 1 = \deg(\mathfrak{m}) - 1. \qquad (10)$$

The ideal $\mathfrak{c}_Q$ is nothing other than the *conductor* of $\mathcal{O}_Q$ into $\mathcal{O}'_Q$.

Examples. a) $\mathfrak{m} = 2P$. The map $p : X \to X_\mathfrak{m}$ being bijective, we can identify X and $X_\mathfrak{m}$. For $Q \neq P$, $\mathcal{O}'_Q = \mathcal{O}_Q$ and $\mathcal{O}'_P$ is the subring of $\mathcal{O}_P$ formed by the functions whose derivative vanishes at P. The completion of $\mathcal{O}'_P$ is thus the subring of $k[[t]]$ generated by t^2 and t^3. The curve $X_\mathfrak{m}$ is analytically isomorphic at P to the curve $y^2 - x^3 = 0$; one says that it has an *ordinary cusp*.

b) $\mathfrak{m} = P_1 + P_2$, with $P_1 \neq P_2$. Here $X_\mathfrak{m}$ is obtained from X by *identifying* P_1 and P_2; the local ring $\mathcal{O}'_Q$ of the point Q thus obtained is formed by rational functions on X which are regular at P_1 and P_2 and which take the same values there. The curve is analytically isomorphic at Q to the (reducible) curve $xy = 0$; one says that it has a *double point with distinct tangent directions* or an *ordinary double point*.

§2. Riemann-Roch theorems

5. Notations

Beyond the hypotheses of no. 2, we will suppose that the curve X (thus also the curve X') is *complete*. We then know that X is a *projective* variety (cf. chap. II, no. 1); moreover, the same argument shows that X' is also a projective variety, but we will not need this fact.

We denote by g the *genus* of X and we put

$$\delta = \sum_{Q \in S'} \delta_Q \qquad (11)$$

$$\pi = g + \delta. \qquad (12)$$

Let D be a divisor on X prime to S; it defines a sheaf $\mathcal{L}(D)$ on X, cf. II.6. Since $X - S$ is biregularly isomorphic to $X' - S'$, we can transport this sheaf to $X' - S'$ and complete it by $\mathcal{O}'_Q$ for $Q \in S'$. Thus we get a subsheaf $\mathcal{L}'(D)$ of the constant sheaf K on X', and by definition

$$\mathcal{L}'(D)_Q = \begin{cases} \mathcal{O}'_Q & \text{if } Q \in S' \\ \mathcal{L}(D)_Q & \text{if } Q \notin S'. \end{cases} \qquad (13)$$

By analogy with the case of a non-singular curve, we put

$$L'(D) = H^0(X', \mathcal{L}'(D)), \quad I'(D) = H^1(X', \mathcal{L}'(D))$$
$$l'(D) = \dim L'(D), \qquad i'(D) = \dim I'(D).$$

From the fact that the sheaf $\mathcal{L}'(D)$ is locally isomorphic to $\mathcal{O}'$, it is a coherent sheaf, and these cohomology groups are *finite dimensional* (this also follows from the proof of thm. 1 below).

In the particular case where X' is the curve $X_\mathfrak{m}$ associated to a modulus $\mathfrak{m}$ (cf. no. 4), we will write $L_\mathfrak{m}(D)$, $I_\mathfrak{m}(D)$, $l_\mathfrak{m}(D)$, and $i_\mathfrak{m}(D)$ in place of $L'(D), \ldots$ etc.

6. The Riemann-Roch theorem (first form)

Theorem 1. *For every divisor D prime to S,*

$$l'(D) - i'(D) = \deg(D) + 1 - \pi. \tag{16}$$

PROOF. Put $\chi'(D) = l'(D) - i'(D)$. The argument of chap. II, no. 4 (based on the exact sequence of cohomology) shows that

$$\chi'(D + P) = \chi'(D) + 1$$

for every $P \in X' - S'$. As the same formula holds for the expression $\deg(D) + 1 - \pi$, we are reduced to proving thm. 1 in the particular case where $D = 0$.

In this case, we have $\mathcal{L}'(D) = \mathcal{O}'$, and everything comes down to showing that the Euler-Poincaré characteristic $\chi(X', \mathcal{O}')$ of $\mathcal{O}'$ is equal to $1 - \pi$. But the sheaf $\mathcal{O}'$ is a subsheaf of $\mathcal{O}$ which is the image of the sheaf of local rings of X (cf. no. 1). From this, we deduce the exact sequence

$$0 \to \mathcal{O}' \to \mathcal{O} \to \mathcal{O}/\mathcal{O}' \to 0,$$

whence

$$\chi(X', \mathcal{O}') = \chi(X', \mathcal{O}) - \chi(X', \mathcal{O}/\mathcal{O}').$$

The sheaf $\mathcal{O}/\mathcal{O}'$ is concentrated on the finite set S'; it follows that

$$\chi(X', \mathcal{O}/\mathcal{O}') = \dim H^0(X', \mathcal{O}/\mathcal{O}') = \sum_{Q \in S'} \mathcal{O}_Q/\mathcal{O}'_Q = \delta.$$

In view of the definition of π, everything thus comes down to proving that

$$\chi(X', \mathcal{O}) = 1 - g.$$

But since g is the genus of X, we have $\chi(X, \mathcal{O}_X) = 1 - g$, cf. chap. II. Theorem 1 will thus be proved when we have established the following result:

Lemma 1. *If an algebraic variety X is the normalization of an algebraic variety X', then $H^q(X,\mathcal{O}) = H^q(X',\mathcal{O})$ for all $q \geq 0$.*

PROOF. For $q = 0$, $H^q(X',\mathcal{O})$ is equal to the set of rational functions $f \in k(X)$ which belong to all the $\mathcal{O}_Q$, $Q \in X'$. As $\mathcal{O}_Q = \bigcap_{P \to Q} \mathcal{O}_P$, it is also the set of $f \in k(X)$ which belong to all the $\mathcal{O}_P$, $P \in X$, whence the equality to prove:

$$H^0(X',\mathcal{O}) = H^0(X,\mathcal{O}). \tag{17}$$

Now let $\mathfrak{U}' = \{U_i'\}$ be a cover of X' by affine opens, and let $\mathfrak{U}$ be the cover of X formed by the $U_i = p^{-1}(U_i')$. In view of the construction of X, the U_i are affine opens and we have

$$H^q(X',\mathcal{O}) = H^q(\mathfrak{U}',\mathcal{O}) \quad \text{and} \quad H^q(X,\mathcal{O}) = H^q(\mathfrak{U},\mathcal{O}).$$

But applying (17) to the varieties $U_{i_0}' \cap U_{i_1}' \cap \cdots \cap U_{i_q}'$ and to their normalizations $U_{i_0} \cap U_{i_1} \cap \cdots \cap U_{i_q}$ we see that the canonical homomorphism $C(\mathfrak{U}',\mathcal{O}) \to C(\mathfrak{U},\mathcal{O})$ is bijective (we denote by $C(\mathfrak{U},\mathcal{O})$ the complex associated to the cover $\mathfrak{U}$ and the sheaf $\mathcal{O}$, cf. FAC, no. 18); it is thus the same for the homomorphisms $H^q(\mathfrak{U}',\mathcal{O}) \to H^q(\mathfrak{U},\mathcal{O})$, as was to be proved. □

Corollary. $\pi = i'(0) = \dim H^1(X',\mathcal{O}')$.

PROOF. We apply formula (16) to the divisor $D = 0$, taking into account the fact that $l'(0) = 1$. □

Example. Let k' be a subfield of k, and let F' be a "function field in one variable" over k'; we mean by this that F' is a regular extension of k' (in the sense of Weil [87], chap. I) of transcendence degree 1 over k'. A projective algebraic curve X' corresponds to this field; it is defined over k', irreducible and k'-normal. By extension of scalars to k, one can also consider X' as a curve defined over k, which is no longer necessarily normal (that is to say non-singular). The preceding corollary, together with an argument using répartitions analogous to that of chap. II, no. 5, shows that the integer π attached to the curve X' coincides with the *genus* of the extension F'/k' (in the sense of Chevalley [15]). The relation $\pi = g + \delta$, with $\delta \geq 0$ makes evident the *decrease of the genus* (ibid., chap. V, §6).

7. Application to the computation of the genus of an algebraic curve

Suppose that the curve X' is embedded in a projective space $\mathbf{P}_r$. As with any projective variety, its *arithmetic genus* $p_a(X')$ is defined. Recall (cf. for example Zariski [102]) that, if $\chi(X')$ is the constant term of the Hilbert polynomial of X' in $\mathbf{P}_r$, by definition

$$1 - p_a(X') = \chi(X').$$

Proposition 3. $p_a(X') = \pi = g + \delta$.

PROOF. Indeed, one knows (FAC, no. 80) that, for every projective variety X', the integer $\chi(X')$ is equal to the alternating sum of the dimensions of the $H^q(X', \mathcal{O}')$. Here we have

$$\dim H^0(X', \mathcal{O}') = 1$$

since X' is connected, and $\dim H^1(X', \mathcal{O}') = \pi$, as we have seen; the $H^q(X', \mathcal{O}')$, $q \geq 2$, are zero. Thus we get $\chi(X') = 1 - \pi$, whence proposition 3 by comparing with the formula defining $p_a(X')$. □

This proposition is interesting because it reduces the computation of the genus g to the easier (see below) computation of the arithmetic genus π and to that of the integer δ, which is purely local. If X' has only "ordinary" singularities (double points with distinct tangent directions and ordinary cusps, cf. no. 4), δ is simply the *number* of these singular points.

We treat for example the case where X' is a *complete intersection* in $\mathbf{P}_r$, of $r - 1$ hypersurfaces of degrees $a_1, \ldots, a_{r-1}$ (this means that the ideal defined by X' in $k[X_0, \ldots, X_r]$ is generated by $r - 1$ homogeneous polynomials of degrees $a_1, \ldots, a_{r-1}$). The computation of the cohomology group $H^1(X', \mathcal{O}')$ presents no difficulties (cf. FAC, no. 78, where the case of a complete intersection of arbitrary dimension is treated) and one finds

$$\pi = \frac{1}{2}a_1 a_2 \ldots a_{r-1}a + 1 \quad \text{with} \quad a = \sum a_i - r - 1. \tag{18}$$

When $r = 2$, this is the *Plücker formula* giving the genus of a plane curve of degree d:

$$g = \frac{1}{2}d(d - 3) + 1 - \delta. \tag{19}$$

8. Genus of a curve on a surface

Let V be a projective non-singular surface and let X' be a curve on V. When V is the plane $\mathbf{P}_2$, the Plücker formula shows that $p_a(X')$ depends only on a "numerical" invariant of X', its degree. We are going to see that this is a general fact.

First we make precise some notations (they are essentially those of Zariski [102], [103]):

If D is a divisor on V, one writes $D \sim 0$, and one says that D is *linearly equivalent to* 0, if D is equal to the divisor (φ) of a rational function φ on V.

One denotes by K the divisor of a non-zero rational differential form of degree 2; it is the *canonical divisor* of V, and it is defined up to linear equivalence.

If D_1 and D_2 are two divisors, one can associate to them an integer, written $D_1.D_2$, characterized by the following properties: $D_1.D_2$ is bilinear, is zero if $D_1 \sim 0$ or $D_2 \sim 0$, and coincides with the degree of the intersection cycle $D_1._V D_2$ when this is defined (i.e., when D_1 and D_2 have no irreducible component in common). One has $D_1.D_2 = D_2.D_1$.

Lemma 2. *If X is a non-singular curve of genus g on V, then*

$$2g - 2 = X.(X + K). \tag{20}$$

PROOF. We can find a differential form ω of degree 2 whose divisor K contains X with multiplicity -1; put $K = -X + D$, with X not figuring in D. Since ω admits X as polar variety of multiplicity 1, the *residue* $\mathrm{Res}_X(\omega)$ of ω along X is well defined: if t is a rational function vanishing along X with multiplicity 1, we can write ω in the form $\omega = dt/t \wedge \alpha$, and $\mathrm{Res}_X(\omega)$ is the restriction to X of the differential form α, which is of degree 1. One checks that $\mathrm{Res}_X(\omega)$ does not depend on the choice of t. This definition shows that the divisor K_X on X of the form $\mathrm{Res}_X(\omega)$ is equal to the intersection cycle $X._V D$, and as $\deg(K_X) = 2g - 2$ (chap. II, no. 9) we deduce that $2g - 2 = X.D = X.(X + K)$, which proves the lemma. □

Now if D is any divisor on V, we write $\mathcal{L}(D)$ for the sheaf of germs of functions whose divisor is locally $\geq -D$; if $D \sim D'$, the sheaf $\mathcal{L}(D)$ is isomorphic to the sheaf $\mathcal{L}(D')$.

On the other hand, for every coherent algebraic sheaf $\mathcal{F}$ on V, we put $\chi(V, \mathcal{F}) = \sum(-1)^i \dim H^i(V, \mathcal{F})$, cf. FAC, no. 79. We write $\chi(V)$ in place of $\chi(V, \mathcal{O}_V)$. With the classical notations (cf. Zariski [**102**], [**103**]), one has $\chi(V) = 1 + p_a(V)$.

Proposition 4. *For every divisor D on V,*

$$\chi(V, \mathcal{L}(D)) = \chi(V) + \frac{1}{2}D.(D - K). \tag{21}$$

PROOF. Suppose at first that $D = -X$, with X non-singular; the sheaf $\mathcal{L}(D)$ is nothing other than the sheaf of ideals defined by X, whence the exact sequence

$$0 \to \mathcal{L}(D) \to \mathcal{O}_V \to \mathcal{O}_X \to 0.$$

Passing to Euler-Poincaré characteristics, we get

$$\chi(V, \mathcal{L}(D)) = \chi(V) - \chi(X).$$

As $\chi(X) = 1 - g = -\frac{1}{2}X.(X + K)$, we indeed find the formula (21).

Now we are going to reduce the general case to the particular case we have just treated. Let E be a hyperplane section of V; applying the results of FAC, no. 66 to the sheaf $\mathcal{L}(-D)$ (or arguing directly), we see that the complete linear series $|-D + nE|$ defines a biregular embedding of V into a projective space, provided that the integer n is large enough. Taking a "general" hyperplane section in this embedding, we get an irreducible non-singular curve X_n such that $-D = nE \sim X_n$, that is to say $D - nE \sim -X_n$. Formula (21) is valid for $-X_n$, thus also for $D - nE$. But the two sides of this formula are polynomials in n: this is evident for the right hand side; for the left, this follows from an elementary theorem on coherent sheaves (FAC, no. 80, proposition 3). These two polynomials coincide for n large enough, thus for all n, and in particular for $n = 0$, as was to be shown. $\square$

Proposition 5. *Let X' be an irreducible singular curve on V and let $p_a(X') = \pi$ be its arithmetic genus. Then*

$$p_a(X') = 1 + \frac{1}{2}X'.(X' + K). \tag{22}$$

PROOF. Here again, there is an exact sequence of sheaves

$$0 \to \mathcal{L}(-X') \to \mathcal{O}_V \to \mathcal{O}_{X'} \to 0,$$

whence the equality $\chi(V, \mathcal{L}(-X')) = \chi(V) - \chi(X')$. Applying (21) and taking into account that $\chi(X') = 1 - p_a(X')$, we find formula (22). $\square$

Examples. 1) When $V = \mathbf{P}_2$, the canonical class K is equal to $-3E$, where E denotes a line; if d is the degree of the curve X', then $X' \sim dE$, whence $X'.X' = d^2$ and $X'.K = -3d$. Formula (22) then gives again the Plücker formula.

2) When V is a *quadric* (in other words the product of two projective lines), the divisor class group admits as a basis two generators E_1 and E_2 of the different rulings, and $K = -2E_1 - 2E_2$. Putting $X'.E_1 = d_1$ and $X'.E_2 = d_2$, formula (22) gives the *formula of C. Segre*:

$$\pi = d_1 d_2 - d_1 - d_2 + 1. \tag{23}$$

One has an analogous formula whenever one has determined a *basis* for the group of divisor classes of V (for numerical equivalence).

Remarks. 1) For every divisor D on V, put $p_a(D) = 1 + \frac{1}{2}D.(D + K)$; the integer $p_a(D)$ is called the *virtual arithmetic genus* of the divisor D, cf. Zariski [102].

2) Put $l(D) = \dim H^0(V, \mathcal{L}(D))$; according to the duality theorem (see chap. II, no. 10, as well as [74] and the report of Zariski [103]), one has $\dim H^2(V, \mathcal{L}(D)) = l(K - D)$. As $\sup(D) = \dim H^1(V, \mathcal{L}(D))$ is always ≥ 0, we see that formula (21) implies the *Riemann-Roch inequality for surfaces*:

$$l(D) + l(K - D) \geq 1 + p_a(V) + \frac{1}{2}D.(D - K). \tag{24}$$

The Riemann-Roch theorem proper (in the form of Hirzebruch [35] in the classical case, and of Grothendieck [29] in the general case) is more precise than the formula (21): it also says that $\chi(V)$ is equal to $\frac{1}{12}(K.K + K_0)$, where K_0 denotes the degree of the "canonical class" of dimension 0 of V.

§3. Differentials on a singular curve

9. Regular differentials on X'

We keep the hypotheses and notations of nos. 2 and 5. Let ω be a differential form on the curve X, the normalization of X', and let $Q \in S'$. One says that ω is *regular* at Q if

$$\sum_{P \to Q} \mathrm{Res}_P(f\omega) = 0 \quad \text{for all } f \in \mathcal{O}'_Q. \tag{25}$$

The set of regular differentials at the point Q will be denoted $\underline{\Omega}'_Q$; it is a $\mathcal{O}'_Q$-submodule of the vector space $D_k(K)$ of all differentials. If we put

$$\underline{\Omega}_Q = \bigcap_{P \to Q} \underline{\Omega}_P, \tag{26}$$

evidently $\underline{\Omega}_Q \subset \underline{\Omega}'_Q$, and one checks immediately that the vector spaces $\mathcal{O}_Q/\mathcal{O}'_Q$ and $\underline{\Omega}'_Q/\underline{\Omega}_Q$ are put in duality by the bilinear form $\sum \mathrm{Res}_P(f\omega)$ figuring in (25).

When $X' = X'_{\mathfrak{m}}$, where $\mathfrak{m} = \sum n_P P$ is a modulus on X, condition (25) is equivalent to the conditions

$$\sum_{P \in S} \mathrm{Res}_P(\omega) = 0 \quad \text{and} \quad v_P(\omega) \geq -n_P \text{ for all } P \in S. \tag{27}$$

We return to the general case. If $Q \notin S'$, put $\underline{\Omega}'_Q = \underline{\Omega}_Q$, which makes sense since Q can be identified with a point of X. The $\underline{\Omega}'_Q$ form a subsheaf $\underline{\Omega}'$ of the constant sheaf $D_k(K)$. It is a coherent sheaf (for $\underline{\Omega}_Q$ and $\underline{\Omega}'_Q/\underline{\Omega}_Q$ are coherent); its sections are, by definition, the *everywhere regular* differentials on X'. We are going to give a characterization of them:

Proposition 6. *In order that ω be everywhere regular on X' it is necessary and sufficient that $\mathrm{Tr}_g(\omega) = 0$ for every rational function g on X which is not a p-th power and which belongs to all the local rings $\mathcal{O}'_Q$, $Q \in S'$.*

(For the definition of the trace $\mathrm{Tr}_g(\omega)$, see chap. II, no. 12.)

PROOF. Suppose that ω is everywhere regular on X', and let $g : X \to \Lambda$ be a rational map of X to the projective line Λ. If the differential form $\operatorname{Tr}_g(\omega)$ on Λ were not zero, it would have a pole λ and there would exist a function h, regular on Λ in a neighborhood of λ, such that $\operatorname{Res}_\lambda(h.\operatorname{Tr}_g(\omega)) \neq 0$. According to the trace formula (chap. II, no. 12),

$$\operatorname{Res}_\lambda(h.\operatorname{Tr}_g(\omega)) = \sum_{g(P)=\lambda} \operatorname{Res}_P(h \circ g . \omega). \tag{28}$$

On the right hand side, the terms corresponding to $P \notin S$ are zero because ω and $h \circ g$ are regular at such a point; if g belongs to the intersection of the $\mathcal{O}'_Q$, the same is true of the sum of the other terms, according to the definition of a regular differential. The left hand side being non-zero, this contradiction indeed shows that $\operatorname{Tr}_g(\omega) = 0$.

Conversely, suppose that ω is not everywhere regular on X'; we are going to construct a function $g : X' \to \Lambda$ and a point $\lambda \in \Lambda$ such that $\operatorname{Res}_\lambda(\operatorname{Tr}_g(\omega)) \neq 0$. We denote by $\mathfrak{m}$ a modulus suported on S and larger than the *conductor* $\mathfrak{c}$; if $g \equiv 0 \bmod \mathfrak{m}$ at all the points P mapping to $Q \in S'$, then $g \in \mathcal{O}'_Q$.

First suppose that ω has a pole of order $n \geq 1$ at a point $P_0 \notin S$. Approximation theory for valuations shows the existence of a function g such that $v_{P_0}(g) = n-1$, $g \equiv \mu \bmod \mathfrak{m}$ (with $\mu \neq \lambda = g(P_0)$), and $g(P) \neq \lambda$ at every pole of ω distinct from P_0. We can also require g to have a simple zero at a given point, which implies that g is not a p-th power. Then

$$\operatorname{Res}_\lambda(\operatorname{Tr}_g(\omega)) = \sum_{g(P)=\lambda} \operatorname{Res}_P(g\omega).$$

By construction, all the terms of the sum on the right hand side are zero with the exception of the term $\operatorname{Res}_{P_0}(g\omega)$, which is evidently $\neq 0$; whence the desired result in this case.

Now suppose that ω is not regular at a point $Q \in S'$, that is to say that there exists $f \in \mathcal{O}'_Q$ with $\sum_{P \to Q} \operatorname{Res}_P(f\omega) \neq 0$. Put $f(Q) = \lambda$; for every point P mapping to Q, let n_P be an integer larger than or equal to the coefficient of P in the modulus $\mathfrak{m}$, as well as to $-v_P(\omega)$. Then choose a function g such that $v_P(f - g) \geq n_P$ if $P \to Q$, $g \equiv \mu \bmod \mathfrak{m}$ if $P \not\to Q$ (with $\mu \neq \lambda$), and $g(P) \neq \lambda$ at every pole P of ω not lying in S. In view of the choice of $\mathfrak{m}$ and the n_P, $g \in \mathcal{O}'_R$ for all $R \in S'$; we can also arrange, as above, that g is not a p-th power. Thus

$$\operatorname{Res}_\lambda(\operatorname{Tr}_g(\omega)) = \sum_{g(P)=\lambda} \operatorname{Res}_P(g\omega) = \sum_{P \to Q} \operatorname{Res}_P(g\omega)$$
$$= \sum_{P \to Q} \operatorname{Res}_P(f\omega) \neq 0,$$

as was to be shown. $\qquad\qquad\square$

10. Duality theorem

Let D be a divisor prime to S; we will associate to it a sheaf $\underline{\Omega}'(D)$ by putting

$$\underline{\Omega}'(D)_Q = \begin{cases} \Omega'_Q & \text{if } Q \in S' \\ \underline{\Omega}(D)_Q & \text{if } Q \notin S'. \end{cases}$$

We also put $\Omega'(D) = H^0(X', \underline{\Omega}'(D))$. A differential ω belongs to $\Omega'(D)$ if it is regular at all the points $Q \in S'$ and if it satisfies the conditions $v_P(\omega) \geq v_P(D)$ for all $P \notin S$. When $D = 0$, we recover the *everywhere regular* differentials on X'.

Theorem 2. *For every divisor D prime to S, the vector space $\Omega'(D)$ is canonically isomorphic to the dual of $I'(D) = H^1(X', \mathcal{L}'(D))$.*

Corollary 1. $i'(D) = \dim \Omega'(D)$.

Corollary 2. *The dimension of the vector space of everywhere regular differential forms on X' is equal to π.*

PROOF. To prove theorem 2 we will need to interpret $\Omega'(D)$ and $I'(D)$ in the language of *répartitions*. Let R be the algebra of répartitions on X (cf. chap. II, no. 5). We denote by R' the subset of R formed by répartitions $\{r_P\}$ such that $r_{P_1} = r_{P_2}$ if P_1 and P_2 have the same image in X'; we will denote by $R'(D)$ the subset of R' formed by the répartitions $\{r_P\}$ such that $v_P(r_P) \geq -v_P(D)$ if $P \notin S$ and $r_P \in \mathcal{O}'_Q$ if P is a point of S having image Q in S'.

We know that the space $D_k(K)$ of all differentials is identified with the topological dual of R/K, that is to say with the set of linear forms ω on R which vanish on K, and on a suitable subset $R(\Delta)$ (for the notations, see chap. II, no. 5). To say that such a linear form belongs to $\Omega'(D)$ means that ω vanishes on $R'(D)$. Thus we see that $\Omega'(D)$ is nothing other than the *topological dual* of the vector space $R/(R'(D) + K)$, endowed with the topology defined by the images of the $R(\Delta)$.

Now we pass to $I'(D)$. There is an exact sequence of sheaves on the space X':

$$0 \to \mathcal{L}'(D) \to K \to K/\mathcal{L}'(D) \to 0. \tag{29}$$

Since K is a constant sheaf and X' is irreducible, $H^0(X', K) = K$ and $H^q(X', K) = 0$ for $q \geq 1$. The exact sequence (29) then gives

$$K \to H^0(X', K/\mathcal{L}'(D)) \to H^1(X', \mathcal{L}'(D)) \to 0. \tag{30}$$

The argument made in chap. II for the sheaf $K/\mathcal{L}(D)$ also applies to the sheaf $\mathcal{A}' = K/\mathcal{L}'(D)$ and shows that $H^0(X', \mathcal{A}')$ is identified with the direct sum of the $\mathcal{A}'_Q$ for $Q \in X'$; as this direct sum is visibly isomorphic to $R'/R'(D)$, we conclude that $I'(D) = H^1(X', \mathcal{L}'(D))$ is canonically identified with the quotient $R'/(R'(D) + K)$.

Now put $V = R'/(R'(D) + K)$ and $W = R/(R'(D) + K)$. We have $V \subset W$, and we must show that *the (algebraic) dual of V can be identified with the (topological) dual of W*, in other words that every linear form on V can be extended uniquely to a continuous linear form on W. This will be the case if we check the two following properties:

i) V is dense in W.
ii) V is discrete for the topology induced by W.

Verification of i). We must show that, for every divisor Δ,

$$R' + R(\Delta) = R.$$

Let $r = \{r_P\}$ be an arbitrary element of R. There exists an element $f \in K$ such that $v_P(f - r_P) \geq v_P(\Delta)$ for all $P \in S$; let g be the répartition $f - r$, and decompose g as $g = g' + g''$, where $g'_P = 0$ for $P \in S$ and $g''_P = 0$ for $P \notin S$. We have $g'' \in R(\Delta)$, $g' \in R'$, and $f \in R'$, which shows that $r = f - g' - g''$ indeed belongs to $R' + R(\Delta)$.

Verification of ii). We must show the existence of a divisor Δ such that

$$R' \bigcap [K + R'(D) + R(\Delta)] = K + R'(D). \tag{31}$$

Let n be an integer such that the relation $v_p(f) \geq n$ for $P \to Q$ implies $f \in \mathcal{O}'_Q$; such an integer exists according to no. 2. We choose for Δ the divisor $\Delta = D - \sum_{P \in S} nP$. Let $r = f + r' + s$ be an element of the left hand side of (31), with $r \in R'$, $f \in K$, $r' \in R'(D)$, and $s \in R(\Delta)$. From the fact that $s = r - f - r'$, we have $s \in R'$, in other words s_P belongs to $\mathcal{O}'_Q$, and, since Δ coincides with D outside of S, we have $s \in R'(D)$, which proves (31) and finishes establishing thm. 2.   □

Remark. One can complete thm. 2 by showing that $H^1(X', \Omega')$ is isomorphic to k and that the duality between $H^0(X', \Omega'(D))$ and $H^1(X', \mathcal{L}'(D))$ is given, as in chap. II, by the cup product.

11. The equality $n_Q = 2\delta_Q$

Let $Q \in S'$, and let $\mathfrak{c}_Q$ be the conductor of $\mathcal{O}_Q$ into $\mathcal{O}'_Q$; from the fact that $\mathfrak{c}_Q$ is an ideal of $\mathcal{O}_Q$ there exists a divisor $\sum_{P \to Q} n_P P$ such that $\mathfrak{c}_Q$ is identical to the set of functions f such that $f \equiv 0 \bmod \sum n_P P$. We can without inconvenience identify the ideal $\mathfrak{c}_Q$ with the divisor $\sum n_P P$.

We can give a simple interpretation of $\mathfrak{c}_Q$ by means of differentials. We have seen in fact that Ω'_Q/Ω_Q is the dual of $\mathcal{O}_Q/\mathcal{O}'_Q$; these two $\mathcal{O}'_Q$-modules thus have the same annililator $\mathfrak{c}_Q$. Thus, $f \in \mathfrak{c}_Q$ if and only if $v_P(f\omega) \geq 0$ for all $P \to Q$ and every $\omega \in \Omega'_Q$. This comes down to saying that the integer n_P *is equal to* $\mathrm{Sup}(-v_P(\omega))$ *for* ω *running through* Ω'_Q.

So denote by $n_Q = \sum_{P \to Q} n_P$ the degree of the divisor $\mathfrak{c}_Q$. We are going to compare n_Q and δ_Q:

Proposition 7. $\delta_Q + 1 \leq n_Q \leq 2\delta_Q$ for every $Q \in S'$, and the equality $n_Q = 2\delta_Q$ holds if and only if $\underline{\Omega}'_Q$ is a free $\mathcal{O}'_Q$-module of rank 1.

PROOF. We have $n_Q = \dim \mathcal{O}_Q/\mathfrak{c}_Q$, and the inclusion $k + \mathfrak{c}_Q \subset \mathcal{O}'_Q$ shows that $n_Q \geq \delta_Q + 1$.

On the other hand, we have just seen that, for every $P \to Q$, there exists a differential $\omega_P \in \underline{\Omega}'_Q$ such that $v_P(\omega) = -n_P$. Forming a linear combination of the ω_P, and taking into account the fact that the field k is infinite, we construct a differential $\omega \in \underline{\Omega}'_Q$ such that $v_P(\omega) = -n_P$ for all $P \to Q$. It is clear that $f\omega \in \underline{\Omega}_Q \Longleftrightarrow f \in \mathfrak{c}_Q$, which shows that the map $f \to f\omega$ defines an injection of $\mathcal{O}'_Q/\mathfrak{c}_Q$ into $\underline{\Omega}'_Q/\underline{\Omega}_Q$, whence the inequality $n_Q - \delta_Q \leq \delta_Q$. Furthermore, if $\underline{\Omega}'_Q$ is a free rank-1 $\mathcal{O}'_Q$-module, the differential ω necessarily constitutes a basis and the map above is surjective, which shows that $n_Q - \delta_Q = \delta_Q$. Conversely, if this equality holds, the map is surjective and every differential $\alpha \in \underline{\Omega}'_Q$ is the sum of a differential $f\omega$, $f \in \mathcal{O}'_Q$, and a differential $\beta \in \underline{\Omega}_Q$; as β can itself be written in the form $g\omega$, with $g \in \mathfrak{c}_Q$, we indeed see that ω constitutes a basis of $\underline{\Omega}'_Q$, which finishes the proof. $\square$

If $n_Q = 2\delta_Q$ for every $Q \in S'$, the sheaf $\underline{\Omega}'$ is *locally free* and thus corresponds to a class of divisors K' on X'; the preceding argument shows that $K = K' - \mathfrak{c}$, whence in particular $2g - 2 = \deg(K') - 2\delta$, that is to say

$$\deg(K') = 2\pi - 2. \tag{32}$$

The sheaves $\underline{\Omega}'(D)$ are thus isomorphic to the sheaves $\mathcal{L}(K' - D)$, and we get the Riemann-Roch theorem in its *second form*:

$$l'(D) - l'(K' - D) = \deg(D) + 1 - \pi. \tag{33}$$

12. Complements

In view of the formulas (32) and (33), it is interesting to give conditions under which $n_Q = 2\delta_Q$. This is the case when X' is a *complete intersection* in a projective space (Rosenlicht [63], § 5). It is also the case when X' is *embedded in a non-singular surface* V; indeed, one can prove that, for every $Q \in X'$, the $\mathcal{O}'_Q$-module $\underline{\Omega}'_Q$ is formed by residues on X' of differentials ω on V such that $(\omega) \geq -X'$ at Q (see Samuel [69] and Gorenstein [26]), and it is then clear that $\underline{\Omega}'_Q$ is a free module of rank 1. This characterization of regular differentials also shows that the canonical divisor K' of X' is equal to $X'._V(X' + K_V)$; using (32), we recover the formula (22) giving the arithmetic genus of X'.

These results, whose direct proof is tiresome, are recovered naturally in the setting of Grothendieck's theory [28]. He has shown ([105], VIII) that, if X' is embedded in a non-singular variety Y of any dimension n, there is a canonical isomorphism

$$\underline{\Omega}'_Q = \text{Ext}^{n-1}_{\mathcal{O}_Q(Y)}(\mathcal{O}'_Q, \underline{\Omega}^n_Q(Y)), \tag{34}$$

where $\mathcal{O}_Q(Y)$ and $\underline{\Omega}^n_Q(Y))$ denote respectively the local ring of Y at the point Q and the module of differential forms of degree n on Y which are regular at Q. When X' is a *complete intersection at* Q, one can write down explicitly a free resolution of the $\mathcal{O}_Q(Y)$-module $\mathcal{O}'_Q$, and this resolution shows that $\underline{\Omega}'_Q$ is a free module of rank 1 over $\mathcal{O}'_Q$. Explicitly, one sees that, if $f_1, \ldots, f_{n-1}$ are generators of the ideal of X' in $\mathcal{O}_Q(Y)$ and if $x_1, \ldots, x_n$ form a regular system of parameters of $\mathcal{O}_Q(Y)$, the module $\underline{\Omega}'_Q$ admits for a basis the differential ω obtained by "division" of $dx_1 \wedge \cdots \wedge dx_n$ by $df_1 \wedge \cdots \wedge df_{n-1}$. For $n = 2$, one has only one equation $f(x, y)$, and ω is written in the classical form $\omega = dx/f'_y = -dy/f'_x$.

Bibliographic note

For a long time in the study of singular curves, the principal problem was that of the resolution of singularities. This was obtained, in the last century, using "quadratic" transformations; the reader will find an exposé of this in Severi [79] or Northcott [59], [60], [61]. The method of normalization, introduced by Zariski, is more rapid; however, it gives a less complete result. For example, if one applies, following Jung, quadratic transformations to the branch curve of the projection of a surface V to a plane, one easily ends up with the resolution of singularities of V (in characteristic 0).

In the other direction, the construction of a singular curve starting with a non-singular curve and a semi-local ring in its field of functions is due to Rosenlicht [63], who also treats the case of a reducible curve defined over any base field. Rosenlicht's memoir also contains the Riemann-Roch theorem as well as the theory of regular differentials. These results were known when the curve has only "ordinary" singularities, cf. Severi [81], chap. I.

We mention a recent memoir of Hironaka [34] showing that the integer $\pi = g + \delta$ coincides with the arithmetic genus of the curve; here again, this fact was well-known to the Italian algebraic geometers in the case of ordinary singularities.

The equality $n_Q = 2\delta_Q$ for curves on a surface was proved by Gorenstein [26] and Samuel [69]; the analogous result in analytic geometry can be found in Kodaira [44], [45]. The point of view sketched at the end of no. 12 was pointed out to me orally by Grothendieck; it is developed in Altman-Kleiman [105], chap. VIII and in Hartshorne [115].

CHAPTER V

Generalized Jacobians

This chapter contains the construction and elementary study of the *generalized Jacobians* of an algebraic curve. We will follow closely the paper of Rosenlicht [64] on this subject, itself inspired by Weil's *Variétés abéliennes* [89], where the case of the usual Jacobian is treated. We will make use, as they did, of the method of "generic points". This obliges us to renounce the point of view of the preceding chapters (where all points had their coordinates in a fixed base field), and to adopt that of *Foundations* [87]. It is certain that the generic points could be replaced by divisors on product varieties, after first developing in detail the properties of these divisors (that is to say essentially the cohomology of coherent algebraic sheaves on a product variety); that would take us too far afield.

§1. Construction of generalized Jacobians

1. Divisors rational over a field

In all of this chapter, X denotes a projective algebraic curve, irreducible, non-singular, and defined over a field k. In §§1, 2 and 3 the field k will be assumed to be *algebraically closed*. In conformity with the conventions of *Foundations* [87], we will make a choice of a *universal domain* Ω; recall that this means that Ω is an algebraically closed extension of k of infinite transcendence degree over k. All the fields considered (except function fields) will be subextensions of Ω of finite type over k. By a *point* of X, we mean a point P all of whose coordinates belong to Ω; if we denote by $k(P)$ the field generated over k by these coordinates, $k(P)$ is an extension of finite type over k. One says that P is rational over a field K if $k(P) \subset K$. The hypothesis that k is algebraically closed implies that X has *infinitely many points rational over k*.

We are now going to recall the definition and principal properties of divisors rational over a field; for the proofs, the reader can refer to *Foundations* [87], or to the work of Samuel [71].

Let $D = \sum n_i P_i$ be a divisor on X. Choose an affine model U of X containing all the P_i and defined over the field k. A coordinate ring $A = k[U]$ is associated to the affine curve U; it is a Dedekind ring (it is normal and of dimension 1). Considering U as a curve over Ω, its coordinate ring is $A_\Omega = A \otimes_k \Omega$. The divisor D is defined by an (in general fractional) ideal $\mathfrak{d}$ of A_Ω. An elementary linear algebra argument then shows that there exists a smallest field K such that $\mathfrak{d}$ is of the form $\mathfrak{o} \otimes_K \Omega$, where $\mathfrak{o}$ is a K vector subspace of $A_K = A \otimes_k K$; in other words, K is the smallest field (containing k) such that $\mathfrak{d}$ is generated by polynomials with coefficients in K. One says that K is the *field of rationality* of D, and one writes $K = k(D)$. If L is an arbitrary field, we denote by $L(D)$ the compositum $L.K(D)$. One says that D is *rational over L* if $L(D) = L$, that is to say if $\mathfrak{d}$ can be generated by equations with coefficients in L. This definition shows that, if D_1 and D_2 are rational over L, the same is true of $D_1 - D_2$ and $\mathrm{Sup}(D_1, D_2)$. If φ is a function which is rational over L (that is to say it belongs to the fraction field of the algebra A_L), its divisor (φ) is rational over L. In the case of a divisor reduced to a point, the definition of $k(D)$ coincides with that of $k(P)$ given above.

More generally, in order that a divisor $D = \sum n_\alpha P_\alpha$, $n_\alpha \neq 0$, be rational over a field, it is necessary and sufficient that the following three conditions be satisfied (cf. Weil, Samuel, loc. cit.):

i) $k(P_\alpha) \subset \overline{L}$, the algebraic closure of L (in other words, D should be *algebraic* over L).

ii) $D^\sigma = D$ for every L-automorphism σ of $\overline{L}$ (or of Ω, it is the same according to i)).

iii) The integer n_α is divisible by $[L(P_\alpha) : L]_i$, the inseparable factor of the degree of the extension $L(P_\alpha)/L$.

Let D be an effective divisor of degree n; one can identify D (cf. chap. III, no. 14) with a point $\overline{D}$ of the *symmetric product* $X^{(n)}$. This last is evidently defined over k, and thus one can speak of the field $k(\overline{D})$. This field *coincides* with $k(D)$ according to a result of Chow, cf. Samuel [71], p. 104 (the proofs of Chow and Samuel use the properties of "Chow coordinates", but it is easy to give a direct proof). In fact, we will only use this result in the following particular case, which is in Weil [89], p. 10, and in Lang [52], p. 30:

Lemma 1. *Let K be a field and let $M_1, \ldots, M_n$ be n independent generic points of X over K. If M denotes the divisor $M_1 + \cdots + M_n$, then $K(M) = K(M_1, \ldots, M_n)_s$, denoting by $K(M_1, \ldots, M_n)_s$ the set of elements of the field $K(M_1, \ldots, M_n)$ invariant by all permutations of $[1, n]$.*

(Recall that the phrase "$M_1, \ldots, M_n$ are independent generic points over K" means that the extension $K(M_1, \ldots, M_n)/K$ has transcendence degree n.)

2. Equivalence relation defined by a modulus

From now on, we make the choice of a *modulus* $\mathfrak{m}$ supported on S; we will assume that the points of S are *rational over* k (if k were not algebraically closed, it would be necessary to suppose that the modulus $\mathfrak{m}$ is rational over k as a divisor—the results of the first two §§ are valid without change, those of §3 should be slightly modified—we will return to this in §4).

Let D and D' be two divisors prime to S. We will say that D and D' are $\mathfrak{m}$-*equivalent*, and we will write $D \sim_{\mathfrak{m}} D'$ if there exists a non-zero rational function g satisfying the two conditions:

a) $g \equiv 1 \bmod \mathfrak{m}$ (cf. chap. III, no. 1),
b) $(g) = D' - D$.

Condition a) should be suppressed if $\mathfrak{m} = 0$. In the general case one can replace it by the condition that g be congruent mod $\mathfrak{m}$ to a non-zero constant.

In conformity with the notations of chap. I, we denote by $C_{\mathfrak{m}}$ the group of classes of divisors prime to S modulo $\mathfrak{m}$-equivalence and by $C_{\mathfrak{m}}^0$ the subgroup of $C_{\mathfrak{m}}$ formed by classes of degree 0. We can also interpret $C_{\mathfrak{m}}$ as the group of *classes of line bundles over the singular curve* $X_{\mathfrak{m}}$ *associated to* $\mathfrak{m}$ (chap. IV, no. 4). The correspondence between line bundles and divisors is established as follows: let Q be the unique singular point of $X_{\mathfrak{m}}$ and let E be a line bundle. We can find a rational section s of E which is regular and non-zero at Q. As $X_{\mathfrak{m}} - Q = X - Q$, we can speak of the *divisor* (s) of s, which is prime to S and defined up to $\mathfrak{m}$-equivalence. One then easily checks that $E \to (s)$ defines an isomorphism from the group of line bundles to the group $C_{\mathfrak{m}}$.

Now let D be a divisor and consider the *effective* divisors D' such that $D \sim_{\mathfrak{m}} D'$. As in chap. II in the case $\mathfrak{m} = 0$, this comes down to considering functions g which satisfy $(g) \geq -D$ outside of S and which satisfy a). Such a function g belongs to the local ring $\mathcal{O}'_Q$ of the singular point Q of the singular curve $X_{\mathfrak{m}}$; thus $g \in L_{\mathfrak{m}}(D)$, with the notations of chap. IV, no. 5. Conversely, *if $g \in L_{\mathfrak{m}}(D)$ is not zero on S, the divisor $D' = (g) + D$ is $\mathfrak{m}$-equivalent to D.* As the functions $g \in L_{\mathfrak{m}}(D)$ which are zero on S are nothing other than the functions $g \in L(D - \mathfrak{m})$, we thus finally see that *the* *set of effective divisors D' such that $D' \sim_{\mathfrak{m}} D$ is in bijective correspondence with the projective space associated to the vector space $L_{\mathfrak{m}}(D)$ minus the projective subspace associated to $L(D - \mathfrak{m})$.*

Lemma 2. *Let D be a divisor prime to S and rational over a field K. There exists a basis of $L_{\mathfrak{m}}(D)$ (resp. of $I_{\mathfrak{m}}(D)^*$) formed by functions (resp. differential forms) rational over K.*

(To avoid any confusion with the universal domain, we have denoted by $I_{\mathfrak{m}}(D)^*$ the space of sections of the sheaf $\underline{\Omega}_{\mathfrak{m}}(D)$, cf. chap. IV, §3.)

PROOF. We must prove that belonging to $L_{\mathfrak{m}}(D)$ imposes K-linear conditions on a function $f \in \Omega(X)$. But these conditions are of two sorts: first there is the condition that f be congruent mod $\mathfrak{m}$ to a constant and second that $(f) \geq -D$. The first is evidently K-linear (it is even k-linear); the second is also by the very definition of the field of rationality of a divisor. Whence the desired result for $L_{\mathfrak{m}}(D)$. As for $I_{\mathfrak{m}}(D)^*$, we first reduce to the particular case $\mathfrak{m} = 0$. We next remark that $I(D)$ is isomorphic to $L(\Delta - D)$, where Δ is a canonical divisor. As one can always choose Δ rational over k, we are reduced to the case treated in the first place. $\square$

Corollary. *If there exists only one effective divisor D' such that $D' \sim_{\mathfrak{m}} D$, this divisor is rational over K.*

PROOF. The uniqueness of D' means that dim $L_{\mathfrak{m}}(D) = 1$ and $L(D-\mathfrak{m}) = 0$. According to the preceding lemma, there exists a function $g \in L_{\mathfrak{m}}(D)$ which is not zero and rational over K. As $D' = (g) + D$, it indeed follows that D' is rational over K. $\square$

3. Preliminary lemmas

Let D be a divisor prime to S. We put (cf. chap. IV, no. 5)

$$l_{\mathfrak{m}}(D) = \dim L_{\mathfrak{m}}(D) \quad \text{and} \quad i_{\mathfrak{m}}(D) = \dim I_{\mathfrak{m}}(D).$$

According to the Riemann-Roch theorem,

$$l_{\mathfrak{m}}(D) - i_{\mathfrak{m}}(D) = \deg(D) + 1 - \pi$$

where π denotes the *arithmetic genus* of the singular curve $X_{\mathfrak{m}}$, that is to say

$$\begin{cases} \pi = g & \text{if } \mathfrak{m} = 0 \\ \pi = g + \deg(\mathfrak{m}) - 1 & \text{if } \mathfrak{m} \neq 0. \end{cases}$$

Lemma 3. *Let K be a field, D a divisor rational over K, and P a generic point of X over K. If $i_m(D) > 0$, then*

$$i_m(D + P) = i_m(D) - 1.$$

PROOF. The vector space $I_m(D + P)^*$ is identified with the subspace of $I_m(D)^*$ formed by differentials ω vanishing at the point P. Thus in any case $i_m(D + P) \geq i_m(D) - 1$ and everything comes down to showing that there exists at least one differential $\omega \in I_m(D)^*$ not vanishing at P. But, since $i_m(D)^* > 0$, lemma 2 shows that there exists a non-zero differential form $\omega \in I_m(D)^*$ which is *rational over* K. The set of points where ω vanishes is necessarily algebraic over K, thus does not contain the point P.   □

The following result is fundamental for what follows:

Lemma 4. *Let D be a divisor of degree 0 rational over a field K and let $M_1, \ldots, M_\pi$ be π independent generic points over K. Then*

a) $l_m(D + \sum_{i=1}^{i=\pi} M_i) = 1$.
b) *There exists a unique effective divisor Δ such that $\Delta \sim_m D + \sum_{i=1}^{i=\pi} M_i$.*
c) $K(\Delta) = K(M_1, \ldots, M_\pi)_s$.

PROOF. Since $\deg(D) = 0$, we have $l_m(D) \leq 1$, and the Riemann-Roch theorem shows that $i_m(D) \leq \pi$. Then repeatedly applying lemma 3, we see that $i_m(D + \sum_{i=1}^{i=\pi} M_i) = 0$, and again applying Riemann-Roch, we find a).

Thus there exists a function g, unique up to multiplication by scalars, belonging to $L_m(D + \sum_{i=1}^{i=\pi} M_i)$. When $m = 0$, this proves b); when $m \neq 0$, we must also check that $L(D + \sum_{i=1}^{i=\pi} M_i - m) = 0$, cf. no. 2. So put $A = D + \sum_{i=1}^{i=\pi} M_i - m$ and suppose that $l(A) \geq 1$. The divisor A has degree $g - 1$ and the usual Riemann-Roch formula shows that $i(A) = l(A) \geq 1$. On the other hand, repeatedly applying lemma 3 (with $m = 0$ this time), we find that $i(A) = i(D - m) - \pi$, whence $i(D - m) > \pi$. But the divisor $D - m$ has degree < 0, whence $l(D - m) = 0$ and, applying Riemann-Roch, $i(D - m) = \pi$, which gives a contradiction.

It remains to prove c). According to lemma 2, the function g can be chosen rational over the field $K(D + \sum_{i=1}^{i=\pi} M_i)$, a field which is equal to $K(M_1, \ldots, M_\pi)_s$ according to lemma 1. As Δ is equal to $(g) + D + \sum_{i=1}^{i=\pi} M_i$, it is rational over $K(M_1, \ldots, M_\pi)_s$. On the other hand, applying a) to $D = 0$, we get $l_m(\sum_{i=1}^{i=\pi} M_i) = 1$, which shows that the divisor $\sum_{i=1}^{i=\pi} M_i$ is the unique divisor ≥ 0 m-equivalent to $\Delta - D$. The argument given above then shows that $K(\sum_{i=1}^{i=\pi} M_i)$ is contained in the field $K(\Delta - D) = K(\Delta)$, whence finally the equality c).   □

4. Composition law on the symmetric product $X^{(\pi)}$

Let $Y = X^{(\pi)}$ be the π-uple symmetric product of X. We are going to endow Y with a rational composition law which will make it a "birational group".

As we have already indicated, every effective divisor of degree π $M = M_1 + \cdots + M_\pi$ can be identified with a point of Y. When the points M_i are generic and independent over a field K, we get a point $M \in Y$ which is a generic point of Y over K, and we have seen that the field $K(M)$ is equal to the field of the *point* M over Y.

We are now going to choose once and for all a point $P_0 \in X$ which is rational over K and not in S; this point will serve as a sort of *origin* for the group laws that we will construct.

Lemma 5. *Let M and N be two independent generic points of Y over a field K. Then there exists a unique divisor R such that $R \sim_\mathfrak{m} M + N - \pi P_0$ and we have*

$$K(M, N) = K(R, M) = K(R, N).$$

PROOF. Put $M = \sum_{i=1}^{i=\pi} M_i$; the points M_i are independent generic points over the field $K(N)$, which is also $K(N - \pi P_0)$ since P_0 is rational over K. Lemma 4 then shows the existence and uniqueness of a divisor R such that

$$R \sim_\mathfrak{m} M + N - \pi P_0.$$

We also see that $K(R, N) = K(M, N)$, whence, exchanging the roles of M and N, $K(R, M) = K(M, N)$, as was to be shown. $\square$

Proposition 1. *There exists a unique rational composition law $F : Y \times Y \to X$ on the variety $Y = X^{(\pi)}$ which is defined over k and is such that, if M and N are two independent generic points of Y over a field K, $F(M, N)$ is the point R defined by lemma 5.*

Furthermore, this composition law makes Y a "birational group" (in other words, it is a normal composition law in the sense of Weil [89], § V).

PROOF. Let M and N be two independent generic points of the variety Y over the field k, and let R be the point of Y defined by lemma 5. From the fact that $k(R) \subset k(M, N)$ there exists a unique rational map $F : Y \times Y \to Y$ which is defined over k and which maps (M, N) to R. Now let M' and N' be two independent generic points of Y over a field K; a fortiori, M' and N' are independent over k. Thus there exists a k-automorphism σ of the universal domain Ω such that $M' = M^\sigma$ and $N' = N^\sigma$. From the fact that F is defined over k, we have

$$F(M^\sigma, N^\sigma) = F(M, N)^\sigma, \quad \text{whence} \quad F(M', N') = R^\sigma.$$

But by hypothesis there exists a function g with $g \equiv 1 \bmod \mathfrak{m}$ such that

$$(g) = M + N - \pi P_0 - R.$$

Applying σ, we deduce

$$(g^\sigma) = M' + N' - \pi P_0 - R^\sigma,$$

with $g^\sigma \equiv 1 \bmod \mathfrak{m}$ (from the fact that $\mathfrak{m}$ is rational over k). This shows that $F(M', N') = R^\sigma$ is nothing other than the divisor associated to (M', N') by lemma 5, which proves the first part of the proposition.

It remains to see that F is a normal law of composition. We must first check that F is *associative for generic points*, in other words that

$$F(M, F(M', M'')) = F(F(M, M'), M'')$$

when M, M' and M'' are three independent generic points. As each of the two sides is the unique effective divisor $\mathfrak{m}$-equivalent to $M+M'+M''-2\pi P_0$, it is clear. Next we must check that, putting $R = F(M, N)$, we have

$$k(M, N) = k(R, M) = k(R, N),$$

which is nothing other than the second assertion of lemma 5. This completes the proof.   □

Note that the composition law F is *commutative*.

5. Passage from a birational group to an algebraic group

We just constructed a "birational group" whose composition law is defined over k. According to a result of Weil ([89], §V, reconsidered and completed in [93]), such a group is birationally isomorphic to a true algebraic group (where the composition law and the inverse map are regular maps, and not just rational). Furthermore, this group, as well as the isomorphism, can be defined over k. We are going to limit ourselves to a rapid indication of the steps of the proof, refering to Weil, loc. cit., for the details.

First we observe that the algebraic group sought for, if it exists, is *unique*. This comes down to saying that every birational isomorphism between two algebraic groups G_1 and G_2 is necessarily biregular, which is a particular case of the following result:

Lemma 6. *Every rational map $f : G_1 \to G_2$ which is a homomorphism for generic points is everywhere regular.*

PROOF. The hypothesis means that there exists a non-empty open U of G_1 such that f is regular on U, and that $f(xy) = f(x)f(y)$ if x, y, $xy \in U$. Fixing x and varying y, we deduce that f is regular on the open xU; as the xU, $x \in U$, cover G_1, the map f is indeed everywhere regular.   □

The *existence* of the algebraic group G sought for is more difficult to establish. First one proves:

Lemma 7. *For every birational group Y defined over a field k, there exists an algebraic group G defined over an extension K/k which is birationally isomorphic over K to Y.*

(Cf. [89], §V, thm. 15, as well as [93], no. 6.)

One begins by constructing an open Y' of Y on which the composition law has sufficient regularity properties (a "group chunk"). One next defines G by glueing several copies of Y' by means of generic translations; it is these translations which oblige one to enlarge the ground field.

Once G is constructed over the extension K/k, one can "descend" its base field to k:

Lemma 8. *For every birational group Y defined over k there exists an algebraic group G_0 which is defined over k and birationally isomorphic over k to Y.*

One uses the theorems of "descent of the base field" taking into account that the extension K/k can be chosen separable (even, in fact, regular); for these descent theorems, see Weil [95].

In the particular case where k is algebraically closed, one can give a direct construction of G_0, cf. Rosenlicht [64], thm. 4. This will be the only case we will need.

6. Construction of the Jacobian $J_\mathfrak{m}$

Combining proposition 1 and lemma 8, we get:

Proposition 2. *There exists an algebraic group $J_\mathfrak{m}$ and a birational map $\varphi : X^{(\pi)} \to J_\mathfrak{m}$ defined over k such that if M and N are two independent generic points of $X^{(\pi)}$,*

$$\varphi(M) + \varphi(N) = \varphi(M * N)$$

where $M * N$ denotes the divisor R of lemma 5.

Furthermore, these properties define φ uniquely up to a unique isomorphism.

The group $J_\mathfrak{m}$ is called the *generalized Jacobian* of X (relative to the modulus $\mathfrak{m}$).

We will study the map φ in more detail in §2. Note for the moment that, since φ is defined over k, $\varphi(M)$ makes sense for every generic point of $X^{(\pi)}$. On the other hand, the composition $M * N$ is defined if M and N are *independent and generic* on $X^{(\pi)}$. More generally, let $M_1, \ldots, M_r$ be r independent generic points of $X^{(\pi)}$; lemma 5 shows that one can define by induction the composition $M_1 * M_2 * \cdots * M_r$ and that it is a generic point of $X^{(\pi)}$. On this subject we have:

Lemma 9. *Let $M_1, \ldots, M_r$ (resp. $N_1, \ldots, N_r$) be r independent generic points of $X^{(\pi)}$. The following three conditions are equivalent:*

a) $M_1 * \cdots * M_r = N_1 * \cdots * N_r$
b) $M_1 + \cdots + M_r \sim_{\mathrm{m}} N_1 + \cdots + N_r$
c) $\varphi(M_1) + \cdots + \varphi(M_r) = \varphi(N_1) + \cdots + \varphi(N_r)$.

PROOF. According to lemma 5, the composition $M_1 * \cdots * M_r$ is the unique effective divisor m-equivalent to $M_1 + \cdots + M_r - (r-1)\pi P_0$, whence a)$\Longleftrightarrow$b). As for the equivalence a)$\Longleftrightarrow$c), it follows from the fact that φ is a bijective homomorphism on generic points.   □

§2. Universal character of generalized Jacobians

In this §, we are going to show that the Jacobian J_{m} defined in §1 does have the universal property announced in thm. 2 of chap. I.

7. A homomorphism from the group of divisors of X to J_{m}

Let D be a divisor on X prime to S. We propose to associate to it an element $\theta(D)$ of J_{m}.

Let K be a field containing $k(D)$ and let $M = \sum_{i=1}^{i=\pi} M_i$ be a generic point of $X^{(\pi)}$ over K. According to lemma 4, there exists a unique divisor $N \in X^{(\pi)}$ such that

$$N \sim_{\mathrm{m}} D - (\deg\ D) P_0 + M$$

and we have $K(N) = K(M)$, which shows that N is generic over K. The images $\varphi(M)$ and $\varphi(N)$ of M and N by $\varphi : X^{(\pi)} \to J_{\mathrm{m}}$ are well-determined elements of J_{m}. We put

$$\theta_M(D) = \varphi(N) - \varphi(M).$$

This definition is legitimate, for it is clear that N does not change if K is made smaller, and the difference $\varphi(N) - \varphi(M)$ is indeed independent of K.

Lemma 10. *Let K be a field and let D and D' be two divisors rational over K; put $D'' = D + D'$. If M, M', and M'' are three independent generic points over K, then*

$$\theta_{M''}(D'') = \theta_M(D) + \theta_{M'}(D').$$

PROOF. Denote by N, N', and N'' the points associated respectively to D, D' and D'' by the procedure above. We must show the relation

$$\varphi(N'') - \varphi(M'') = \varphi(N) - \varphi(M) + \varphi(N') - \varphi(M'),$$

which can also be written

$$\varphi(N) + \varphi(N') + \varphi(M'') = \varphi(N'') + \varphi(M) + \varphi(M').$$

As $K(N) = K(M)$ and $K(N') = K(M')$, the three generic points N, N', and M'' are independent, and the same is true of the three others N'', M, and M'. According to lemma 9, the formula to be proved is equivalent to the following, which is clear:

$$N + N' + M'' \sim_{\mathrm{m}} N'' + M + M'.$$

$\square$

Lemma 11. $\theta_M(D)$ *does not depend on* M.

PROOF. First remark that $\theta_M(0) = 0$, for then $N = M$. Applying lemma 10 with $D' = 0$, we deduce that

$$\theta_M(D) = \theta_{M''}(D)$$

when M and M'' are generic and independent.

Now if M' is another generic point over K, we can always find a third generic point M'' which is simultaneously independent of M and M', and, by virtue of the proceeding, $\theta_M(D) = \theta_{M''}(D) = \theta_{M'}(D)$, as was to be shown. $\square$

From now on, we will write $\theta(D)$ in place of $\theta_M(D)$.

Lemma 12. *If* $P = \sum_{i=1}^{i=\pi} P_i$ *is a generic point of* $X^{(\pi)}$, *then* $\theta(P) = \varphi(P)$.

PROOF. Let M be a generic point of $X^{(\pi)}$ independent of P. Forming the divisor $N \in X^{(\pi)}$ such that

$$N \sim_{\mathrm{m}} P + M - \pi P_0$$

we recognize the divisor denoted $P * M$ in prop. 2. By definition,

$$\theta(P) = \theta_M(P) = \varphi(N) - \varphi(M) = \varphi(P * M) - \varphi(M) = \varphi(P),$$

by virtue of proposition 2. $\square$

Lemma 13. *If* D *is rational over a field* K, *the point* $\theta(D) \in J_{\mathrm{m}}$ *is rational over* K.

PROOF. The construction of $\theta(D)$ in the form $\varphi(N) - \varphi(M)$ shows that $\varphi(D)$ is rational over the field $K(M) = K(N)$. As this holds for any M generic over K, we conclude that $k(\theta(D))$ is contained in the intersection of the fields $K(M)$, an intersection which is equal to K. $\square$

Proposition 3. *The map θ is a homomorphism from the group of divisors prime to S onto the group $J_{\mathfrak{m}}$. Its kernel is formed by the divisors $\mathfrak{m}$-equivalent to an integral multiple of P_0.*

PROOF. Lemma 10 shows that θ is a homomorphism. Its image is a subgroup of $J_{\mathfrak{m}}$ which, according to lemma 12, contains all the generic points of $J_{\mathfrak{m}}$. This subgroup is thus equal to $J_{\mathfrak{m}}$ (indeed, every point of an algebraic group is the product of two generic points). Thus θ is surjective.

In order that a divisor D be such that $\theta(D) = 0$, it is necessary and sufficient that

$$\varphi(M) = \varphi(N)$$

(the notations M and N being as in the beginning of this no.). By virtue of lemma 9 (or even of the very definition of φ), this means that $M \sim_{\mathfrak{m}} N$, that is to say that $D \sim_{\mathfrak{m}} (\deg\ D)P_0$.   □

8. The canonical map from X to $J_{\mathfrak{m}}$

First we are going to prove an auxiliary result on coverings:

Lemma 14. *Let X and X' be two curves (complete and non-singular, as always) and let $g : X \to X'$ be a separable covering of degree $n + 1$. For every $P \in X$, the divisor*

$$g^{-1}(g(P))$$

is of the form $P + H_P$, where H_P is an effective divisor of degree n. If H_P is identified with a point of the symmetric product $X^{(n)}$, the map $P \to H_P$ is a regular map from X to $X^{(n)}$.

PROOF. We will use the same method as in no. 13 of chap. III. Let Y be a Galois covering of X' dominating X, let $\mathfrak{g}$ be the Galois group of $Y \to X'$, and let $\mathfrak{h} \subset \mathfrak{g}$ be the Galois group of $Y \to X$. Choose representatives σ_i, $i = 1, \ldots, n+1$, of the classes of $\mathfrak{g}$ mod $\mathfrak{h}$ and let σ_{n+1} be the representative of the class $\mathfrak{h}$. If we denote by π the projection $Y \to X$, we know (cf. III-13) that $g^{-1}(g(P)) = \sum \pi \circ \sigma_i(Q)$, where Q denotes a point of Y mapping to P. We conclude that

$$H_P = \sum_{i=1}^{i=n} \pi \circ \sigma_i(Q).$$

Then we define a regular map $h : Y \to X^n$ by putting

$$h(Q) = (\pi \circ \sigma_1(Q), \ldots, \pi \circ \sigma_n(Q)).$$

By composing h with the canonical map $X^n \to X^{(n)}$, we get a regular map

$$h' : Y \to X^{(n)}.$$

The formula above shows that h' defines by passage to the quotient a map $h'' : X \to X^{(n)}$ which is clearly regular and which is none other than the map $P \to H_P$. ☐

Remark. One can consider lemma 14 as a particular case of the "subtraction" theorem which can be stated as follows:

Let T and X be two algebraic varieties and suppose that to every $t \in T$ are associated three positive cycles of dimension zero in X, say H_t, H'_t, and H''_t, with

$$H_t = H'_t + H''_t.$$

Let n, n', and n'' be the degrees of these cycles, which are assumed to be independent of t. We say that the family H_t is *regular* if the corresponding map $H : T \to X^{(n)}$ is regular, and similarly for H' and H''. Then, *if two of the three families H_t, H'_t, and H''_t are regular, so is the third*. The proof reduces to an exercise in symmetric functions.

We now return to the Jacobian $J_{\mathfrak{m}}$ and to the map θ. This map is defined for all divisors of X prime to S, thus in particular for the *points* $P \in X - S$. We propose to prove:

Proposition 4. *The map* $\theta : X - S \to J_{\mathfrak{m}}$ *is a rational map which is everywhere regular and defined over k.*

We rely on the following lemma:

Lemma 15. *Let M be a generic point of $X^{(\pi)}$ over a field K. There exists a rational map $\theta' : X \to J_{\mathfrak{m}}$ defined over the field $K(M)$ which is regular at all the points of $X - S$ rational over K and coincides at these points with θ.*

Admit for a moment this lemma. As M and K vary the map θ' does not change, since it coincides with θ in all of the points of $X - S$ which are rational over k. One can thus speak of *the map* θ' and, as every point of X is rational over a suitable field K, we see that θ' is regular on $X - S$ and coincides there with θ. Finally, θ' is defined over the intersection of the fields $K(M)$, an intersection which is nothing other than k.

It remains to prove the lemma. We suppose that $\pi \neq 0$, otherwise there is nothing to prove, the group $J_{\mathfrak{m}}$ being reduced to the identity element.

Let P_1 be a point of $X - S$ distinct from the point P_0 chosen in no. 4 and rational over k. According to lemma 4, there exists a divisor $N \in X^{(\pi)}$ such that

$$N \sim_{\mathfrak{m}} -P_1 + P_0 + M$$

and $K(N) = K(M)$.

Let g be a rational function such that

$$(g) = N + P_1 - P_0 - M$$
$$g \equiv 1 \bmod \mathfrak{m}.$$

From the fact that $l_{\mathfrak{m}}(-P_1 + P_0 + M) = 1$, this function g is unique, and it is defined over $K(M)$, cf. the proof of lemma 4. The divisor $(g)_{\infty}$ of *poles* of g is less than or equal to the divisor $P_0 + M$. We are going to show that it is *equal* to $P_0 + M$.

First of all, the point P_0 is a pole of g, since it is distinct from P_1 and does not figure in N (N is a generic divisor over K). Thus we must show that, if $M = \sum_{i=1}^{i=\pi} M_i$, all the M_i are poles of g. Let us show this for one of them, say M_π. If this were not the case, we would have

$$g \in L_{\mathfrak{m}}\left(P_0 - P_1 + \sum_{i<\pi} M_i\right) \quad \text{and} \quad l_{\mathfrak{m}}\left(P_0 - P_1 + \sum_{i<\pi} M_i\right) \geq 1$$

whence, by Riemann-Roch,

$$i_{\mathfrak{m}}\left(P_0 - P_1 + \sum_{i<\pi} M_i\right) \geq 1$$

and, by virtue of lemma 3, $i_{\mathfrak{m}}(P_0 - P_1) \geq \pi$. Again applying Riemann-Roch, this would imply $l_{\mathfrak{m}}(P_0 - P_1) \geq 1$, whence the existence of a function h such that

$$h \equiv 1 \bmod \mathfrak{m} \quad \text{and} \quad (h) = P_0 - P_1.$$

This means that h is a biregular isomorphism of X to the projective line Λ, and the relation $h \equiv 1 \bmod \mathfrak{m}$ is only possible if $\mathfrak{m} = 0$; but in this case, we would have $\pi = g + \delta = 0$, a case which we have excluded.

Thus the divisor of poles of g is indeed $P_0 + M$ and that of zeros is $P_1 + N$. We can consider $g : X \to \Lambda$ as a covering of degree $\pi + 1$. From the fact that P_0 is a simple pole, g is not a p-th power and this covering is separable. Thus we can apply lemma 14 to it: putting $g^{-1}(g(P)) = P + H_P$, the map $P \to H_P$ is a regular map $s : X \to X^{(\pi)}$. Furthermore, this map is defined over $K(M)$, as is easily seen. Then we put

$$\theta'(P) = \varphi(M) - \varphi(H_P).$$

The map $\theta' : X \to J_{\mathfrak{m}}$ is a rational map defined over $K(M)$. We are going to see that it has all the requisite properties.

Thus we suppose that P is rational over K and is not contained in S, and we let $a = g(P)$. Suppose $P \neq P_0$, in which case $a \neq \infty$. Then $(g - a) \geq P - P_0 - M$, which shows that $g - a$ is the unique function (up to a constant factor) belonging to $L_{\mathfrak{m}}(P_0 - P + M)$, cf. lemma 4. According to the proof of this lemma, $g - a$ is not zero on S and, if $g' = (g - a)/(1 - a)$, then $g' \equiv 1 \bmod \mathfrak{m}$. We have

$$(g') = g^{-1}(a) - (g)_{\infty} = P + H_P - P_0 - M,$$

whence

$$H_P \sim_{\mathrm{m}} -P + P_0 + M.$$

Thus, the divisor associated by lemma 4 to $-P + P_0 + M$ is nothing other than H_P, which shows that H_P is generic over K and $\varphi(H_P)$ makes sense. Furthermore, by the very definition of θ

$$\theta(-P) = \varphi(H_P) - \varphi(M),$$

whence

$$\theta(P) = \theta'(P).$$

When $P = P_0$ (the case we have excluded), this formula is trivial.

Finally, θ' is indeed regular at P, for, up to a translation of J_{m}, it is the composition of $s : X \to X^{(\pi)}$, which is regular (lemma 14), with $-\varphi : X^{(\pi)} \to J_{\mathrm{m}}$, which is regular at $s(P) = H_P$ since this is a generic point of $X^{(\pi)}$. This finishes the proof of lemma 15 and at the same time that of proposition 4. $\qquad\square$

9. The universal property of the Jacobians J_{m}

Lemma 16. *The canonical extension of $\theta : X - S \to J_{\mathrm{m}}$ to the symmetric product $X^{(\pi)}$ coincides with the rational map φ of proposition 2.*

PROOF. Let $S\theta$ be the canonical extension of θ to the symmetric product $(X - S)^{(\pi)}$. It is a regular map from $(X - S)^{(\pi)}$ to J_{m}, and one can thus consider it as a *rational* map from $X^{(\pi)}$ to J_{m}. To show that it coincides with φ, it suffices to prove that $S\theta(M) = \varphi(M)$ when $M = \sum_{i=1}^{i=\pi} M_i$ is a *generic* point of $X^{(\pi)}$. But by definition

$$S\theta(M) = \sum_{i=1}^{i=\pi} \theta(M_i) = \theta(M)$$

and lemma 12 shows that $\theta(M) = \varphi(M)$. $\qquad\square$

There is thus no difference between the maps φ and θ. To unify the notations, we choose the notation φ (or φ_{m} when we want to make explicit m) to denote the canonical map $X \to J_{\mathrm{m}}$ as well as its extension to the group of divisors prime to S. Observe that all of this construction depends on the choice of an "origin" point P_0; moreover, $\varphi(P_0) = 0$. Changing P_0 only changes φ by a translation, as we will see a little later.

The following theorem recapitulates the properties of the map $\varphi : X \to J_{\mathrm{m}}$.

Theorem 1.

a) *The map* $\varphi_{\mathfrak{m}} : X \to J_{\mathfrak{m}}$ *is a rational map defined over k and regular at every point of $X - S$.*

b) *The extension of $\varphi_{\mathfrak{m}}$ to the divisors prime to S defines, by passage to the quotient, an isomorphism from the group $C_{\mathfrak{m}}^0$ of classes of divisors of degree 0 (with respect to $\mathfrak{m}$-equivalence) to the group $J_{\mathfrak{m}}$.*

c) *The extension of $\varphi_{\mathfrak{m}}$ to the symmetric product $X^{(\pi)}$ is a birational map from $X^{(\pi)}$ to $J_{\mathfrak{m}}$.*

It is clear that, conversely, properties a), b), c), and the normalization condition $\varphi_{\mathfrak{m}}(P_0) = 0$ characterize $J_{\mathfrak{m}}$ and the map $\varphi_{\mathfrak{m}}$ uniquely.

We now prove that $J_{\mathfrak{m}}$ has the universal property announced in chap. I.

Theorem 2. *Let $f : X \to G$ be a rational map from X to a commutative group G admitting $\mathfrak{m}$ for a modulus and put $g_0 = f(P_0)$. Then there exists a unique algebraic homomorphism $F : J_{\mathfrak{m}} \to G$ such that $f = F \circ \varphi_{\mathfrak{m}} + g_0$.*

PROOF. After effecting a translation on f, we can assume that $g_0 = 0$. If D is a divisor of degree 0 and prime to S on X, the element $f(D) \in G$ only depends on the $\mathfrak{m}$-equivalence class of D. By passage to the quotient, we get a homomorphism $C_{\mathfrak{m}}' \to G$ from the group of these classes to G. According to thm. 1 b), there thus exists unique a homomorphism $F : J_{\mathfrak{m}} \to G$ such that $f = F \circ \varphi_{\mathfrak{m}}$ for all points $P \in X$ (or for every divisor, it comes to the same thing). It remains to see that F is an *algebraic* homomorphism. But let Sf be the extension of f to $X^{(\pi)}$; it is a rational map from $X^{(\pi)}$ to G which is regular on $(X - S)^{(\pi)}$ by the very definition of a symmetric product. In view of the definition of F, $Sf = F \circ \varphi$, denoting by φ the canonical map $X^{(\pi)} \to J_{\mathfrak{m}}$. But φ is biregular on a non- empty open U of $J_{\mathfrak{m}}$. It follows that F coincides on U with the regular map $Sf \circ \varphi^{-1}$ and by translation, we deduce that F is everywhere regular.   □

Remark. The proof shows at the same time that, if f is defined over a field K, the same is true of Sf, thus of F.

Corollary 1. *In the construction of the pair $(\varphi_{\mathfrak{m}}, J_{\mathfrak{m}})$ changing P_0 does not change $J_{\mathfrak{m}}$ and changes $\varphi_{\mathfrak{m}}$ only by a translation.*

PROOF. We apply theorem 2 to the modified map $\varphi_{\mathfrak{m}}' : X \to J_{\mathfrak{m}}$, which gives a homomorphism $F : J_{\mathfrak{m}} \to J_{\mathfrak{m}}'$. We also get a homomorphism $F' : J_{\mathfrak{m}}' \to J_{\mathfrak{m}}$ and the uniqueness property of theorem 2 shows that $F \circ F' = 1$ and $F' \circ F = 1$. We can thus identify $J_{\mathfrak{m}}$ and $J_{\mathfrak{m}}'$ and then $\varphi_{\mathfrak{m}}' = \varphi_{\mathfrak{m}} + \varphi_{\mathfrak{m}}'(P_0)$.   □

Corollary 2. *Every rational map from X to a commutative group can be factored by means of a suitable map $\varphi_{\mathfrak{m}}$.*

PROOF. Indeed, according to theorem 1 of chap. III, such a map always has at least one modulus.   □

10. Invariant differential forms on J_m

Since J_m is a commutative algebraic group of dimension π, the invariant differential forms on J_m form a vector space of dimension π. If ω is such a form, $\alpha = \varphi^*(\omega)$ is a differential form on X, which is clearly regular outside of S. More precisely:

Proposition 5. *The map $\omega \to \varphi^*(\omega)$ is a bijection between the set of invariant differential forms on J_m and the set of differential forms α on X satisfying $(\alpha) \geq -m$.*

PROOF. First suppose that $\varphi^*(\omega) = 0$. Let $g : X^\pi \to J_m$ be the map $(x_1, \ldots, x_\pi) \to \sum \varphi(x_i)$ and let $h_i : X^\pi \to J_m$ be the map $(x_1, \ldots, x_\pi) \to \varphi(x_i)$. From the fact that $g = \sum h_i$, we have $g^*(\omega) = \sum h_i^*(\omega)$ (cf. chap. III, prop. 16), whence $g^*(\omega) = 0$. But the map g factors as $X^\pi \to X^{(\pi)} \to J_m$ and the two partial maps $X^\pi \to X^{(\pi)}$ and $X^{(\pi)} \to J_m$ are generically surjective and separable. [Recall that a rational map $f : X \to Y$, where X and Y are irreducible, is called *generically surjective* if $f(X)$ is dense in Y; the field $k(Y)$ then is identified with a subfield of $k(X)$, and if the extension $k(X)/k(Y)$ is separable (resp. purely inseparable, primary), one says that f is separable (resp. purely inseparable, primary).] The map g is itself also generically surjective and separable, whence the fact that $g^*(\omega) = 0$ implies $\omega = 0$, by virtue of the characterization of separable extensions by differentials (see, for example [11], exposé 13). The map $\omega \to \varphi^*(\omega)$ is thus *injective*.

Let us then write $\Omega(-m)$ for the vector space of differential forms α on X such that $(\alpha) \geq -m$. The Riemann-Roch theorem shows that $\dim \Omega(-m) = \pi$, which is also the dimension of the space of invariant differential forms on J_m. Thus it will suffice to show that $\varphi^*(\omega) \in \Omega(-m)$ for every invariant differential ω on J_m.

The differentials $\alpha \in \Omega(-m)$ have no poles outside the set S. The residue formula thus shows that $\sum_{P \in S} \text{Res}_P(\alpha) = 0$, and comparing with chap. IV, no. 9, we see that $\Omega(-m)$ *is nothing other than the set of everywhere regular differentials on X_m*. In order to check that $\alpha = \varphi^*(\omega)$ belongs to $\Omega(-m)$, it thus suffices to show (loc. cit., prop. 6) that $\text{Tr}_g(\alpha) = 0$ for every rational function g on X such that $g \equiv 0 \mod m$ and which is not a p-th power.

But, let $h = \text{Tr}_g(\varphi)$ be the map of the projective line Λ to the group J_m defined as in chap. III, no. 2. According to lemma 4 of chap. III, no. 6, $\text{Tr}_g(\alpha) = h^*(\omega)$. On the other hand, since m is a modulus for $\varphi : X \to J_m$, the map h is constant (chap. III, no. 5, prop. 9). It follows that $h^*(\omega) = 0$, which finishes the proof.   □

Corollary 1. *Let J be the usual Jacobian of X. The map $\omega \to \varphi^*(\omega)$ is a bijection from the set of invariant differentials on J to the set of differentials of the first kind on X.*

PROOF. This is the particular case $\mathfrak{m} = 0$. □

Corollary 2. *If $\deg(\mathfrak{m}) \geq 2$, the map $\varphi_\mathfrak{m}$ is not regular at any point of S.*

PROOF. Let $P \in S$ and let n_P be the coefficient of P in $\mathfrak{m}$. In view of the hypothesis $\deg(\mathfrak{m}) \geq 2$ we have $\mathfrak{m} - P > 0$. The Riemann-Roch theorem then shows that

$$\dim \ \Omega(-\mathfrak{m} + P) = \pi - 1$$

and there exists a differential $\alpha \in \Omega(-\mathfrak{m})$ which does not belong to $\Omega(-\mathfrak{m} + P)$, that is to say which has a pole of order exactly n_P at P. According to prop. 5, such a differential is of the form $\varphi_\mathfrak{m}^*(\omega)$, where ω is regular on $J_\mathfrak{m}$; it is thus impossible that $\varphi_\mathfrak{m}$ is regular at P. □

Remark. The case $\deg(\mathfrak{m}) = 1$ is a trivial case: the curve $X_\mathfrak{m}$ is reduced to X and $\mathfrak{m}$-equivalence is ordinary linear equivalence. The corresponding Jacobian is nothing other than the usual Jacobian.

§3. Structure of the Jacobians $J_\mathfrak{m}$

11. The usual Jacobian

When $\mathfrak{m} = 0$, the generalized Jacobian $J_\mathfrak{m}$ reduces to the usual Jacobian.

According to thm. 1, the canonical map $\varphi : X \to J$ (defined up to a translation) is everywhere regular; the same is thus true of its extension to the symmetric product $X^{(g)}$. As $X^{(g)} \to J$ is birational, the image of $X^{(g)}$ is dense in J and, since $X^{(g)}$ is a complete variety, this image is closed and equal to J. Thus we see that J is the image of a complete variety, thus is *complete*: it is thus an *Abelian variety*.

Furthermore, the map $\varphi : X \to J$ is *universal* for rational maps of X to Abelain varieties. Indeed, if $f : X \to A$ is such a map, we know (since A is a complete variety) that f is everywhere regular, and theorem 1 of chap. III shows that f admits the modulus $\mathfrak{m} = 0$. Theorem 2 then implies that f factors as $f = F \circ \varphi$, where $F : J \to A$ is an "affine" homomorphism (that is to say a homomorphism in the usual sense, followed by a translation). One expresses this property by saying that J is the *Albanese variety* of X. One knows (see Lang [52]) that such a variety exists for every algebraic variety X; we will come back to this later.

12. Relations between Jacobians $J_\mathfrak{m}$

Let $\mathfrak{m}$ and $\mathfrak{m}'$ be two moduli with $\mathfrak{m} \geq \mathfrak{m}'$. To these two moduli (and to the choice of an origin P_0) are associated Jacobians $J_\mathfrak{m}$ and $J_{\mathfrak{m}'}$ and maps $\varphi_\mathfrak{m}$ and $\varphi_{\mathfrak{m}'}$ that we are going to compare:

Proposition 6. *There exists a unique homomorphism $F : J_\mathfrak{m} \to J_{\mathfrak{m}'}$ such that $\varphi_{\mathfrak{m}'} = F \circ \varphi_\mathfrak{m}$. This homomorphism is surjective, separable, and its kernel is a connected subgroup $H_{\mathfrak{m}/\mathfrak{m}'}$ of $J_\mathfrak{m}$.*

PROOF. The map $\varphi_{\mathfrak{m}'} : X \to J_{\mathfrak{m}'}$ admits $\mathfrak{m}'$ as a modulus, thus *a fortiori* also $\mathfrak{m}$. As $\varphi_{\mathfrak{m}'}(P_0) = 0$, theorem 2 shows the existence and the uniqueness of F.

Now let $s : X^{(\pi')} \to X^{(\pi)}$ be the map obtained by passage to the quotient from the map

$$(M_1, \ldots, M_{\pi'}) \to (M_1, \ldots, M_{\pi'}, P_0, \ldots, P_0).$$

Since $J_\mathfrak{m}$ is birationally isomorphic to $X^{(\pi)}$ and $J_{\mathfrak{m}'}$ to $X^{(\pi')}$, the map s canonically defines a rational map

$$s' : J_{\mathfrak{m}'} \to J_\mathfrak{m}.$$

Furthermore, it is evident that $F \circ s' = 1$, that is to say that s' is a "rational section" for the projection $F : J_\mathfrak{m} \to J_{\mathfrak{m}'}$. All the properties of F announced in the proposition then follow from the existence of this section, with the added fact that $J_\mathfrak{m}$ is *birationally isomorphic* to the product $J_{\mathfrak{m}'} \times H_{\mathfrak{m}/\mathfrak{m}'}$. $\square$

The preceding proposition essentially shows that the Jacobians $J_\mathfrak{m}$ form a *projective system* of groups. In fact, this is not exactly correct, since one cannot choose the *same* origin point P_0 for *all* the moduli $\mathfrak{m}$ at the same time. All that one can say is that the $J_\mathfrak{m}$ form a projective system of *principal homogeneous spaces* (cf. no. 21, as well as chap. VI).

Note that $J_{\mathfrak{m}'}$ is identified with the quotient $J_\mathfrak{m}/H_{\mathfrak{m}/\mathfrak{m}'}$. The knowledge of the Jacobians associated to sufficiently large moduli $\mathfrak{m}$ suffices to determine all the Jacobians (including the more general Jacobians J_0 that Rosenlicht has associated to any singular curve having X as normalization).

13. Relation between $J_\mathfrak{m}$ and J

Put $\mathfrak{m}' = 0$ in prop. 6. We get the fact that $J_\mathfrak{m}$ is an *extension* of the usual Jacobian J by a connected group that we will denote $L_\mathfrak{m}$ and whose structure we are going to study.

From the set-theoretic point of view, this offers no difficulties: according to thm. 1 a point $d \in J_\mathfrak{m}$ corresponds to the $\mathfrak{m}$-equivalence class of a divisor

D prime to S which we can assume has degree 0. The image of d in J is 0 if and only if D is linearly equivalent to 0, that is to say if there exists a rational function g such that $D = (g)$. From the fact that D is prime to S, this function is a *unit* at every point $P \in S$, in other words $v_P(g) = 0$.

Conversely, every function g satisfying this condition defines a divisor $D = (g)$ whose class d belongs to the kernel $L_\mathfrak{m}$ in question. We have $d = 0$ in $J_\mathfrak{m}$ if D is of the form (h) with $h \equiv 1 \bmod \mathfrak{m}$, which implies $g = \lambda.h$ where λ is a constant $\neq 0$.

So let us denote by U_P the multiplicative group of functions f such that $v_P(f) = 0$ and by $U_P^{(n)}$ the subgroup of U_P formed by the functions f such that $v_P(1 - f) \geq n$. The function g defines for each P an element $g_P \in U_P/U_P^{(n_P)}$, where n_P denotes the coefficient of P in $\mathfrak{m}$. Conversely, one knows that to every system of g_P corresponds a function g. We are thus led to form the product group

$$R_\mathfrak{m} = \prod_{P \in S} U_P/U_P^{(n_P)}.$$

Each of the factor groups contains as a subgroup the group $\mathbf{G}_m$ of constants; we denote by Δ the "diagonal" group formed by the $(\lambda, \lambda, \ldots, \lambda)$, with $\lambda \in \mathbf{G}_m$, and we put

$$H_\mathfrak{m} = R_\mathfrak{m}/\Delta.$$

It then follows from the proceeding that:

Proposition 7. *The map $g \to (g)$ defines by passage to the quotient a bijective homomorphism from the group $H_\mathfrak{m}$ to the group $L_\mathfrak{m}$, the kernel of the canonical homomorphism $J_\mathfrak{m} \to J$.*

It remains to describe the structure of *algebraic group* on $L_\mathfrak{m}$, and it is this that we are going to do now.

14. Algebraic structure on the local groups $U/U^{(n)}$

Let U be the multiplicative group of formal series $f(t)$ such that $v(f) = 0$, and let $U^{(n)}$ be the subgroup of U formed by those for which $v(1 - f) \geq n$. An element $f \in U^{(n)}$ can thus be written

$$f = 1 + a_n t^n + \cdots .$$

It is clear that the quotient group $U/U^{(n)}$ admits as a system of representatives the polynomials

$$f = a_0 + a_1 t + \cdots + a_{n-1} t^{n-1}, \quad a_0 \neq 0.$$

Thus $U/U^{(n)}$ can be considered as an open subset of the affine space of dimension n and endowed with the corresponding algebraic structure.

Lemma 17. *The preceding algebraic structure is compatible with the group structure of $U/U^{(n)}$. Furthermore, it is independent of the choice of uniformizer t.*

PROOF. If $f = a_0 + a_1 t + \cdots + a_{n-1} t^{n-1}$ and $g = b_0 + b_1 t + \cdots + b_{n-1} t^{n-1}$ are two elements of $U/U^{(n)}$, their product h has $c_0 + c_1 t + \cdots + c_{n-1} t^{n-1}$ for a representative, with

$$c_i = \sum_{r+s=i} a_r b_s.$$

The composition law is thus given by polynomial formulas, which shows that it is everywhere regular. Similarly one checks the regularity of the inverse and the regularity of the operation defined by a "change of variable"

$$t' = \alpha_1 t + \cdots, \quad \alpha_1 \neq 0. \qquad \square$$

Let $V_{(n)}$ be the group $U^{(1)}/U^{(n)}$, the subgroup of $U/U^{(n)}$ formed by the expansions

$$f = 1 + a_1 t + \cdots + a_{n-1} t^{n-1}.$$

Lemma 18. *The group $U/U^{(n)}$ is isomorphic to the product of the group $\mathbf{G}_m$ by the group $V_{(n)}$.*

PROOF. Indeed, every function f can be written uniquely as a product of a constant $a_0 \neq 0$ by a function $g \in V_{(n)}$. Further, this decomposition is compatible with the algebraic structure of $U/U^{(n)}$, since a_0 is a regular function on $U/U^{(n)}$. $\qquad \square$

The group $V_{(n)}$ is a variety biregularly isomorphic to the affine space of dimension $n - 1$. More precisely:

Lemma 19. *For every integer i, $1 \leq i \leq n-1$, let g_i be a formal series of order i. Then every element of $V_{(n)}$ can be written uniquely in the form of a product*

$$g = (1 + a_1 g_1) \cdots (1 + a_{n-1} g_{n-1})$$

where the a_i are constants. The map which sends g to $(a_1, \ldots, a_{n-1})$ is a biregular map of $V_{(n)}$ to the affine space of dimension $n - 1$.

PROOF. Let $g = 1 + b_1 t + \cdots + b_{n-1} t^{n-1}$, and let $g_1 = c_1 t + \cdots$. Putting $a_1 = b_1/c_1$, the quotient $g/(1 + a_1 g_1) = h_1$ satisfies $v(1 - h_1) \geq 2$. Next we determine a_2 such that the quotient $h_1/(1 + a_2 g_2) = h_2$ satisfies $v(1 - h_2) \geq 3$, and so on. At each stage, the coefficient a_i and the function h_i are determined by polynomial formulas, which are thus everywhere regular. $\qquad \square$

As an *algebraic group*, $V_{(n)}$ admits a composition sequence formed of factors isomorphic to $\mathbf{G}_a$. Furthermore, we have:

Lemma 20. *The group $V_{(n)}$ is a unipotent group, isomorphic to the group of matrices of the form:*

$$\begin{pmatrix} 1 & a_1 & a_2 & a_3 & \cdots & a_{n-1} \\ 0 & 1 & a_1 & a_2 & \cdots & a_{n-2} \\ 0 & 0 & 1 & a_1 & \cdots & a_{n-3} \\ 0 & 0 & 0 & 1 & \cdots & a_{n-4} \\ \vdots & \vdots & \vdots & \vdots & \ddots & \vdots \\ 0 & 0 & 0 & 0 & \cdots & 1 \end{pmatrix}$$

PROOF. The isomorphism is obtained by making t correspond to the nilpotent Jordan matrix of order n. $\square$

15. Structure of the group $V_{(n)}$ in characteristic zero

When the characteristic of the base field is 0, one can define the *exponential* $\exp(g) = 1 + g + \cdots + g^n/n! + \cdots$ of any formal series g whose order is > 0, and one has the usual formula

$$\exp(g_1 + g_2) = \exp(g_1) \exp(g_2).$$

Proposition 8. *For every integer i, $1 \leq i \leq n-1$, let g_i be a formal series of order i. Every element $g \in V_{(n)}$ can then be written uniquely in the form of a product*

$$g = \exp(a_1 g_1) \cdots \exp(a_{n-1} g_{n-1})$$

where the a_i are constants. The map which sends g to $(a_1, \ldots, a_{n-1})$ is a biregular isomorphism from the algebraic group $V_{(n)}$ to the group $(\mathbf{G}_a)^{n-1}$.

PROOF. The existence and uniqueness of the above decomposition and the fact that it furnishes a biregular map $V_{(n)} \to (\mathbf{G}_a)^{n-1}$ are proved exactly as in the case of lemma 19. The only new point is the fact that the group structure is preserved, which follows from the formula $\exp(g_1 + g_2) = \exp(g_1) . \exp(g_2)$. $\square$

Corollary. *In characteristic zero, the local group $U/U^{(n)}$ is isomorphic to the product $\mathbf{G}_m \times (\mathbf{G}_a)^{n-1}$.*

16. Structure of the group $V_{(n)}$ in characteristic $p > 0$

One can no longer use the exponential series. In its place, we use the series

$$F(t) = \exp\left(-(t + t^p/p + \cdots + t^{p^n}/p^n + \cdots)\right).$$

A simple computation shows that this series (with rational coefficients) can be put in the form of an infinite product

$$F(t) = \prod_{(n,p)=1} (1 - t^n)^{\mu(n)/n},$$

where μ denotes the Möbius function.

This second expression for F makes evident the fact that its coefficients are *p-adic integers*; thus F makes sense in characteristic p.

Now let $\vec{x} = (x_0, x_1, \dots)$ be a *Witt vector* (of finite or infinite length), and consider the product series

$$E(\vec{x}) = F(x_0).F(x_1) \dots$$

Coming back to the "phantom components" $x^{(0)}, x^{(1)}, \dots,$ of $\vec{x}$, we see that

$$E(\vec{x}) = \exp(-x^{(0)} - x^{(1)}/p - x^{(2)}/p^2 - \cdots)$$

and, in view of the definition of addition of Witt vectors, this implies that

$$E(\vec{x} + \vec{y}) = E(\vec{x}).E(\vec{y})$$

if $\vec{x}$ and $\vec{y}$ are of infinite length.

According to the principle of prolongation of identities, this formula remains valid in characteristic p. *For Witt vectors, the function E replaces the exponential; it is the Artin-Hasse exponential*, cf. [1], [24].

If $\vec{x}$ is a Witt vector and if t is a scalar, we denote by $\vec{x}.t$ the product of $\vec{x}$ by the Witt vector $(t, 0, \dots)$. The components of $\vec{x}.t$ are

$$(x_0 t, x_1 t^p, \dots, x_n t^{p^n}, \dots).$$

The series $E(\vec{x}.t)$ is well defined and has value

$$E(\vec{x}.t) = F(x_0 t).F(x_1 t^p) \cdots = (1 - x_0 t + \cdots)(1 - x_1 t^p + \cdots) \cdots.$$

We can now announce the result which, in characteristic p, replaces proposition 8.

Proposition 9. *For every integer i prime to p and $\leq n - 1$, let r_i be the smallest integer r such that $p^r \geq n/i$; choose a formal series g_i of order i. Every element $g \in V_{(n)}$ can then be put uniquely in the form of a product*

$$g = \prod_{\substack{(i,p)=1 \\ 1 \leq i \leq n-1}} E(\vec{a}_i.g_i)$$

where the $\vec{a}_i$ are Witt vectors of length r_i. The map which sends g to the a_i is a biregular isomorphism from the algebraic group $V_{(n)}$ to the product of the Witt groups W_{r_i}.

The proof is essentially the same as that of Lemma 19 and that of prop. 8.

Corollary. *In characteristic $p > 0$, the local group $U/U^{(n)}$ is isomorphic to the product of $\mathbf{G}_m$ by Witt groups.*

17. Relation between J_m and J: determination of the algebraic structure of the group L_m

We return to the hypotheses and notations of no. 13. The group R_m is the product of the local groups $U_P/U_P^{(n_P)}$ and, according to the preceding nos., each of these groups is canonically endowed with an algebraic group structure. The group R_m is thus also an algebraic group. Further, using lemma 18, we have the decomposition

$$R_m = \prod_{P \in S} \mathbf{G}_{m,P} \times \prod_{P \in S} V_{(n_P)}$$

where $\mathbf{G}_{m,P}$ denotes a group isomorphic to the multiplicative group and attached to the point P. (This is a particular case of the decomposition of a commutative linear group as a product of multiplicative groups and a unipotent group.)

The diagonal group Δ of no. 13 is contained in the factor $\prod_{P \in S} \mathbf{G}_{m,P}$; it is even a direct factor, for, if P_1 is a point of S, one can write

$$\prod_{P \in S} \mathbf{G}_{m,P} = \Delta \times \prod_{P \in S - P_1} \mathbf{G}_{m,P}.$$

The quotient group $H_m = R_m/\Delta$ is thus isomorphic to the product of the $\mathbf{G}_{m,P}$, $P \in S - P_1$, and of $V_{(n_P)}$, $P \in S$. We denote by θ the canonical bijection $\theta : H_m \to L_m$. We have:

Theorem 3. *The map $\theta : H_m \to L_m$ is a biregular isomorphism.*

(In other words, the structure of *algebraic group* of L_m is obtained by transport of structure from that of H_m, which we have just determined.)

PROOF. The proof is in several steps:

Lemma 21. *Let g and h be two rational functions on X. For every element λ of the projective line Λ, let $D_\lambda = (g + \lambda h)$ and let T be the subset of Λ formed by the elements $\lambda \in \Lambda$ such that D_λ has a point in common with S. The map $\lambda \to \varphi(D_\lambda)$ is then a regular map from $\Lambda - T$ to L_m.*

PROOF. After changing g and h, we can suppose that $D_\lambda = (g - \lambda)$. Then let $\psi = \mathrm{Tr}_g(\varphi)$; this is a regular map from $\Lambda - T$ to J_m (cf. chap. III), and

$$\varphi(D_\lambda) = \psi(\lambda) - \psi(\infty)$$

which shows that $\lambda \to \varphi(D_\lambda)$ is a regular map from $\Lambda - T$ to J_m. As $\varphi(D_\lambda) \in L_m$ and L_m has the structure induced from that of J_m, the lemma is proved. $\square$

Lemma 22. *The map $\theta : H_{\mathfrak{m}} \to L_{\mathfrak{m}}$ is regular.*

PROOF. It suffices to show that each of the partial maps $\theta : \mathbf{G}_{m,P} \to L_{\mathfrak{m}}$ and $\theta : V_{(n_P)} \to L_{\mathfrak{m}}$ is regular. Let $\lambda \in \mathbf{G}_{m,P}$; by definition, $\theta(\lambda) = \varphi((u_\lambda))$, where u_λ is a rational function on X congruent to λ mod $\mathfrak{m}$ at P and congruent to 1 mod $\mathfrak{m}$ at the points of $S - P$. But let v be a function such that

$$\begin{cases} v \equiv 1 \bmod \mathfrak{m} & \text{at } P \\ v \equiv 0 \bmod \mathfrak{m} & \text{on } S - P. \end{cases}$$

We can take $u_\lambda = (\lambda - 1).v + 1$ and lemma 21 then shows that $\lambda \to \varphi((u_\lambda))$ is indeed a regular map from $\mathbf{G}_{m,P} = \Lambda - \{0, \infty\}$ to $L_{\mathfrak{m}}$.

We argue similarly for $V_{(n_P)}$: for each i, $1 \le i \le n_P - 1$, we choose a function g of order i at P and of order $\ge n_Q$ at $Q \in S - P$. Lemma 21 shows that the map $\lambda_i \to \varphi((1 + \lambda_i g_i))$ is a regular map from $\Lambda - \{\infty\}$ to $L_{\mathfrak{m}}$, whence the result by applying lemma 19. $\qquad\square$

Thus the map $\theta : H_{\mathfrak{m}} \to L_{\mathfrak{m}}$ is a bijective and regular homomorphism. In characteristic 0, these properties imply that it is biregular; it is not the same in characteristic p (one can simply affirm that it is purely inseparable). We must also prove that the tangent map to θ is bijective, or what comes to the same, that $\theta^*(\beta) = 0$ implies $\beta = 0$ if β is an invariant differential 1-form on $L_{\mathfrak{m}}$ (cf. chap. III, no. 11, cor. 2 to prop. 16). But the group $L_{\mathfrak{m}}$ is defined as a subgroup of the Jacobian $J_{\mathfrak{m}}$. Its invariant differentials are induced from those of $J_{\mathfrak{m}}$, a differential ω inducing 0 on $L_{\mathfrak{m}}$ if and only if it comes from a differential on the usual Jacobian J. We are thus reduced to proving this:

Lemma 23. *If ω is an invariant differential form on $J_{\mathfrak{m}}$ not coming from an invariant differential form on J, $\theta^*(\omega) \ne 0$.*

PROOF. Let $\alpha = \varphi^*(\omega)$ be the differential induced on X by ω. By virtue of prop. 5 and its corollary 1, $(\alpha) \ge -\mathfrak{m}$, and α is nothing other than a differential of the first kind. Thus let $P \in S$ be a pole of α and let n be its order; we have $1 \le n \le n_P$. First we suppose that $n \ge 2$ and let g be a rational function of order $n - 1$ at P, of order $\ge n_Q$ at $Q \in S - P$, and not a p-th power. Let $\psi = \mathrm{Tr}_g(\varphi)$; it is a regular map from $\Lambda - \{0\}$ to $J_{\mathfrak{m}}$. On the other hand, the map $\lambda \to 1 + \lambda g$ defines a regular map from $\Lambda - \{\infty\}$ to $U_P / U_P^{(n_P)}$ (cf. lemmas 19 and 22), and by composition, a regular map $h : \Lambda - \{\infty\} \to J_{\mathfrak{m}}$. From the fact that h can be factored by θ, it will suffice to show that $h^*(\omega) \ne 0$. But the maps h and ψ are related by the formula

$$h(\lambda) = \psi(-1/\lambda) - \psi(\infty).$$

It will thus suffice to prove that $\psi^*(\omega) \ne 0$. According to lemma 4 of chap. III, no. 6, $\psi^*(\omega) = \mathrm{Tr}_g(\alpha)$. Denoting the identity map from Λ to Λ

by λ, the trace formula shows that

$$\operatorname{Res}_0(\lambda \operatorname{Tr}_g(\alpha)) = \sum_{g(Q)=0} \operatorname{Res}_Q(g\alpha) = \operatorname{Res}_P(g\alpha)$$

which is $\neq 0$ in view of the hypotheses made on g and α. Thus $\operatorname{Tr}_g(\alpha) \neq 0$ which completes the proof in the case $n \geq 2$. When $n = 1$, we take for g a function $\equiv 1 \bmod \mathfrak{m}$ at P, and $\equiv 0 \bmod \mathfrak{m}$ on $S - P$; one shows that $\operatorname{Res}_1(\lambda \operatorname{Tr}_g(\alpha)) \neq 0$, and finishes in the same fashion. $\qquad\square$

Remark. Thus $J_\mathfrak{m}$ is an "extension" of the usual Jacobian by a linear group $L_\mathfrak{m}$ which we have made explicit. For example, when $\mathfrak{m} = 2P$, $L_\mathfrak{m} = \mathbf{G}_a$; when $\mathfrak{m} = P + Q$, $P \neq Q$, $L_\mathfrak{m} = \mathbf{G}_m$. But the knowledge of J and of $L_\mathfrak{m}$ is evidently not enough to determine $J_\mathfrak{m}$: one must also determine the type of the extension. We will come back to this in chap. VII.

18. Local symbols

Let $g \in U_P$ with $P \in S$; the element g defines by passage to the quotient an element of $U_P/U_P^{(nP)}$, thus also an element of the group $H_\mathfrak{m}$, which we will denote by $\bar{g}$.

Proposition 10. $\theta(\bar{g}) = -(\varphi_\mathfrak{m}, g)_P$.

PROOF. Let g' be a rational function such that

$$\begin{cases} g' \equiv g \bmod \mathfrak{m} & \text{at } P \\ g' \equiv 1 \bmod \mathfrak{m} & \text{on } S - P \end{cases}$$

By definition, $\theta(\bar{g}) = \varphi_\mathfrak{m}((g'))$; on the other hand, properties i) and ii) of local symbols show that

$$(\varphi_\mathfrak{m}, g')_P = (\varphi_\mathfrak{m}, g)_P$$

$$(\varphi_\mathfrak{m}, g')_Q = 0 \quad \text{if } Q \in S - P.$$

Applying properties iii) and iv) of local symbols, we deduce

$$(\varphi_\mathfrak{m}, g)_P = -\sum_{Q \notin S} (\varphi_\mathfrak{m}, g')_Q = -\sum_{Q \notin S} v_Q(g') \varphi_\mathfrak{m}(Q) = -\varphi_\mathfrak{m}((g'))$$

whence the desired result. $\qquad\square$

Thus the map $\theta : H_\mathfrak{m} \to L_\mathfrak{m}$ is nothing other than the combination of the local symbols relative to the points $P \in S$ (up to a change of sign).

We also note that if a map $f : X \to G$ factors as $X \to J_\mathfrak{m} \xrightarrow{F} G$ (cf. thm. 2), we would have $(f, g)_P = F((\varphi_\mathfrak{m}, g)_P)$, whence $(f, g)_P = -F \circ \theta(\bar{g})$. Thus *the knowledge of the local symbols $(f, g)_P$ is equivalent to the* *knowledge of the restriction of F to L_m.* We also obtain the fact that the local symbols are *regular* functions of g with respect to the structure of algebraic group on the groups $U_P/U_P^{(n_P)}$.

19. Complex Case

We suppose that the base field k is the field $\mathbf{C}$ of complex numbers; the algebraic structure of J_m then determines an *analytic* structure on J_m, which makes it a complex Lie group. We propose to determine this group.

We will need two lemmas:

Lemma 24. *Let G be a connected commutative algebraic group, let G' be an analytic group, and let $p : G' \to G$ be a finite connected covering (in the sense of topology) of G which is a homomorphism of analytic groups. Then there exists a unique algebraic group structure on G' compatible with its analytic structure, and such that p is a regular rational map.*

PROOF. Let N be the kernel of p; it is a finite group, and thus there exists an integer n such that $N \subset G'_n$, denoting by G'_n the subgroup of elements of order n of G'. Multiplication by n is a homomorphism of G' into itself whose tangent map is surjective. Thus it makes G' a covering of itself, with kernel the finite group G'_n; as $N \subset G'_n$, this homomorphism defines by passage to the quotient a homomorphism $h : G \to G'$ and the composition $p \circ h$ is multiplication by n in G'. The kernel H of h is finite, and G' is identified with G/H; one can thus endow G' with the *quotient* algebraic group structure. It is immediate that this structure satisfies the imposed conditions and that it is unique. $\qquad\qquad\qquad\qquad\square$

Lemma 25. *The hypotheses being those of the preceding lemma, let $f : X \to G'$ be a continuous map of an irreducible algebraic variety X to the group G'. In order that f be an everywhere regular rational map, it is necessary and sufficient that $p \circ f$ be a regular rational map from X to G.*

PROOF. The necessity is clear. Thus we suppose that $p \circ f = g$ is an everywhere regular rational map; from the fact that $G' \to G$ is a covering, we conclude already that f is *holomorphic*. Let $X' \to X$ be the pull-back by g of the covering (in the algebraic sense) $G' \to G$; by definition, X' is identified with the subvariety of $X \times G'$ formed by the pairs (x, y') such that $g(x) = p(y')$. According to a known result (cf. for example Weil [96], Appendix), the connected components of X' are the same for the usual topology and for the Zariski topology. But the map f defines a holomorphic section s of the covering $X' \to X$ by the formula $s(x) = (x, f(x))$. It follows that $s(X)$ is an *irreducible component* of X'. The graph of s in $X \times X'$ is thus an algebraic subvariety and the projection $s(X) \to X$ is a regular rational map which is an analytic isomorphism; it is thus a biregular isomorphism ([75], prop. 9). Thus the map s is regular, and the same is true of f.   □

Remarks. 1. One can avoid the recourse to [75] by using the fact that the complete local rings of X and of $s(X)$ coincide. If X is normal, one can also invoke Zariski's "main theorem".

2. We have used the fact that an irreducible algebraic variety is connected for the usual topology. In the particular case of curves one can give a very simple proof (Chevalley [15], p. 141). First of all, one can suppose that the curve X is non-singular (since the image of a connected space is connected) and complete (since a connected surface remains connected when one removes a finite set of points). This being the case, suppose that X has at least two connected components X_1 and X_2 and let $P \in X_1$. Applying the Riemann-Roch theorem to the divisor nP, with n large enough, we see that there exists a non-constant function f on X having P as its only pole. The function f thus induces on X_2 an everywhere holomorphic function, which is thus constant in virtue of the maximum principle; this is absurd, for a non-constant rational function only takes each value a finite number of times.

Return now to the Jacobian $J_\mathfrak{m}$. If T denotes its tangent space at the origin, T is the dual of the space of invariant differential forms on $J_\mathfrak{m}$, itself canonically isomorphic to $\Omega(-\mathfrak{m})$, cf. prop. 5. Thus T is intrinsically determined by X and $\mathfrak{m}$. Further, the exponential map $\exp : T \to J_\mathfrak{m}$ makes T a covering of $J_\mathfrak{m}$ whose kernel we will denote by $\Gamma_\mathfrak{m}$; it is a discrete subgroup of T.

One can describe the map $\varphi_\mathfrak{m} : X - S \to J_\mathfrak{m}$ in the following manner: let P_0 be an origin fixed once and for all, and let $P \in X - S$. Choose a path γ in $X - S$ from P_0 to P. For every differential form $\omega \in \Omega(-\mathfrak{m})$, the integral $\int_{P_0}^{P} \omega$ (taken along γ) depends linearly on ω; one can thus identify it with an element $\theta(\gamma)$ of T. By definition,

$$\langle \theta(\gamma), \omega \rangle = \int_\gamma \omega.$$

The element $\theta(\gamma)$ only depends on the *homotopy class* of γ on $X - S$ and thus defines a map

$$\theta : \widetilde{X} - \widetilde{S} \to T$$

of the universal covering $\widetilde{X} - \widetilde{S}$ of $X - S$ to T. By it very construction, the map $\exp \circ \theta : \widetilde{X} - \widetilde{S} \to J_\mathfrak{m}$ has the same derivative as the composed map $\widetilde{X} - \widetilde{S} \to X - S \to J_\mathfrak{m}$. As both map P_0 to 0, they coincide.

This shows in particular that, if γ is a *cycle* of degree 1, the element $\theta(\gamma)$ belongs to the kernel $\Gamma_\mathfrak{m}$ of $\exp : T \to J_\mathfrak{m}$. Thus we get a canonical homomorphism

$$\theta : H_1(X - S) \to \Gamma_\mathfrak{m},$$

where H_1 denotes the 1-dimensional homology group with integral coefficients.

Proposition 11. *The homomorphism θ is a bijection of $H_1(X - S)$ with Γ_{m}.*

PROOF. Let s be the number of points of S. According to what we have seen in the preceding nos., the group J_{m} is an extension of J by a product of $s-1$ groups $\mathbf{G}_m$ and of some groups $\mathbf{G}_a$. The fundamental group $\pi_1(J_{\mathrm{m}})$ is thus an extension of $\pi_1(J)$ by the group $\mathbf{Z}^{s-1}$; it is a free Abelian group of rank $2g + s - 1$. On the other hand, one knows that $H_1(X)$ is free of rank $2g$, and the exact sequence of homology then shows that $H_1(X - S)$ is free of rank $2g + s - 1$. Since the two groups $\Gamma_{\mathrm{m}} = \pi_1(J)$ and $H_1(X - S)$ are free groups of the same rank, it will suffice to see that θ is *surjective*.

Suppose that this is not so; then there would exist a subgroup Γ' of Γ_{m}, of finite index > 1 in Γ_{m}, containing the image of $H_1(X - S)$ by θ. Let $J' = T/\Gamma'$; it is a finite covering of J_{m}. Furthermore, the hypothesis that Γ' contains $\theta(H_1(X - S))$ shows that the map φ_{m} lifts to a continuous (and even holomorphic) map $\psi : X - S \to J'$. Lemmas 24 and 25 show that J' is canonically endowed with the structure of algebraic group for which the map ψ is rational and regular. Applying prop. 14 of chap. III to ψ, we see that m is a modulus for ψ and according to thm. 2 there exists a homomorphism $F : J_{\mathrm{m}} \to J'$ such that $\psi = F \circ \varphi_{\mathrm{m}}$. The composition of F with the projection $J' \to J_{\mathrm{m}}$ is thus the identity, which is absurd and finishes the proof.   □

Identifying Γ_{m} with $H_1(X - S)$ by means of θ, we have:

Corollary. *The group J_{m} is analytically isomorphic to the quotient of the dual of $\Omega(-\mathrm{m})$ by the discrete subgroup $H_1(X - S)$.*

For the usual Jacobian, this is a well known result.

Remark. In general, the analytic structure of J_{m} *does not determine uniquely its algebraic structure.* Indeed it is easy to give examples of algebraic groups having *non-algebraic analytic automorphisms* (the group $\mathbf{G}_a \times \mathbf{G}_m$) and also examples of algebraic groups which are *analytically isomorphic without being algebraically isomorphic* ($\mathbf{G}_m \times \mathbf{G}_m$ is analytically isomorphic to an extension of an elliptic curve J by a group $\mathbf{G}_a$).

§4. Construction of generalized Jacobians: case of an arbitrary base field

20. Descent of the base field

Let k_1 be an extension of a field k and let V be an algebraic variety defined over k_1 (we also say a k_1-*variety*). "To descend the base field of V from k_1 to k" means to find a k-variety W which is biregularly isomorphic to V over k_1. In other words, one should have a biregular isomorphism $f : W \to V$ defined over k_1.

We suppose from now on that k_1 is a finite *Galois* extension of k; let $\mathfrak{g}$ be its Galois group. If σ is an element of $\mathfrak{g}$, we denote by V^σ the variety obtained from V by means of σ. From the point of view of *Foundations* [87], where V is defined by charts and glueings u_{ij}, the variety V^σ is defined by the same charts and by the glueings u_{ij}^σ; from the point of view of the *schemas* of Chevalley [11], V^σ is defined by the same field and the same places as V, only the structure of k_1-algebra being modified by σ. This also applies to W, but from the fact that W is a k-variety, one can identify W and W^σ. The transform f^σ of f by σ is thus a biregular isomorphism from W to V^σ, and this permits us to put

$$h_\sigma = f^\sigma \circ f^{-1}. \tag{*}$$

The h_σ, $\sigma \in \mathfrak{g}$ are k_1-isomorphisms of V to the V^σ; they satisfy the identity

$$h_{\sigma\tau} = (h_\sigma)^\tau \circ h_\tau. \tag{**}$$

We are thus led to make precise the notion of descent of the base field in the following manner: given k_1-isomorphisms $h_\sigma : V \to V^\sigma$ satisfying (**), we seek a k-variety W and an isomorphism $f : W \to V$ satisfying (*). It is this precise problem that we will consider from now on.

Proposition 12.
a) *If descent of the base field is possible, its solution is unique, up to a k-isomorphism.*
a) *Descent of the base field is possible when the variety V is the union of affine opens U_i defined over k_1 such that*

$$h_\sigma(U_i) \subset U_i^\sigma. \tag{***}$$

PROOF. Let $f : W \to V$ and $f' : W' \to V$ be two solutions of a descent of the base field and let $\varphi = f^{-1} \circ f'$; by hypothesis, φ is a biregular isomorphism, defined over k_1. On the other hand, the formulas $f^\sigma \circ f^{-1} = h_\sigma = f'^\sigma \circ f'^{-1}$ show that $\varphi^\sigma = \varphi$ for all $\sigma \in \mathfrak{g}$, whence the fact that φ is defined over k, which establishes a).

To prove b), we can limit ourselves to the case where the variety V is affine; let A be its coordinate ring, considered as a k_1-algebra. The coordinate ring of V^σ is nothing other than A, considered as k_1-algebra by means of σ, and to give the isomorphism $h_\sigma : V \to V^\sigma$ is equivalent to giving an automorphism $\overline{\sigma}$ of A extending the automorphism σ of k_1. Condition (**) means that $\overline{\sigma\tau} = \overline{\sigma}.\overline{\tau}$, in other words that the group $\mathfrak{g}$ acts on A. Let $B = A^{\mathfrak{g}}$ be the set of elements of A invariant by the actions $\overline{\sigma}$, $\sigma \in \mathfrak{g}$. Since $[k_1 : k] < +\infty$, the ring A is a k-algebra of finite type, and every element of A is integral over B. According to lemma 10 of chap. III, it follows that B is a k-algebra of finite type. Let W be the affine k-variety having B for coordinate ring. To show that W answers the question, it suffices to prove that the algebra $B \otimes_k k_1$ is identified with A. This follows from the following well-known lemma:

Lemma 26. *Let E be a k_1-vector space and suppose given for every $\sigma \in \mathfrak{g}$ a σ-linear bijection $\overline{\sigma}$ of E to itself such that $\overline{\sigma\tau} = \overline{\sigma}.\overline{\tau}$. If F denotes the set of elements of E invariant by the operations $\overline{\sigma}$, $E = F \otimes_k k_1$. (In other words, E has a basis of elements invariant by the $\overline{\sigma}$.)*

PROOF. We recall briefly a proof of this lemma. Put $r = [k_1 : k]$ and let C be the endomorphism algebra of the k-vector space k_1. The scalar multiplications form a subalgebra of C which can be identified with k_1. On the other hand, the k-linear combinations of elements of $\mathfrak{g}$ form another subalgebra D of C. Let $\theta : k_1 \otimes_k D \to C$ be the k-linear map defined by the product. The theorem on the independence of automorphisms shows that θ is injective, and as $k_1 \otimes_k D$ and C both have dimension r^2, we conclude that θ is bijective. This shows that $k_1 \otimes_k D$, endowed with a suitable algebra structure (that of the "crossed product") is a *simple* algebra. But giving the operations $\overline{\sigma}$ endows E with a $k_1 \otimes_k D$-module structure. According to a well known result, it follows that E is the direct sum of simple modules all isomorphic to k_1, which shows that E is of the form $F \otimes_k k_1$, and finishes the proof. (See Bourbaki, *Algèbre*, chap. VIII for more details.)   □

Corollary 1.   *Descent of the base field is possible when the variety V satisfies the following condition:*
(****) *Every finite subset of V formed of points algebraic over k_1 is contained in an affine open which is algebraic over k_1.*

PROOF. We first note that in the condition (****) we can require that the affine open be defined over k_1: it suffices to replace it by the intersection of its conjugates over k_1. To prove that (****) $\Longrightarrow$ (***), let x be a point of V algebraic over k_1, and for each $\sigma \in \mathfrak{g}$ choose an extension of σ to $k_1(x)$ which we again denote by σ. The point x^σ is then well defined and belongs to V^σ. If we put $y_\sigma = h_\sigma^{-1}(x^\sigma)$, condition (****) shows the existence of an affine open U of V containing all the y_σ. Furthermore, according to what was said at the beginning, we can suppose that U is defined over k_1. Then put

$$U' = \bigcap h_\tau(U)^{\tau^{-1}}.$$

The open U' is an affine open of V defined over k_1 and contains the point x because $x^\tau \in h_\tau(U)$. A direct computation shows that $h_\sigma(U') = U'^\sigma$. The opens U' thus have all the required properties and, since they cover the set of points of V algebraic over k_1 they also cover V itself.   □

Corollary 2. *Descent of the base field is possible when V is:*

i) *either a locally closed subvariety of a projective space*
ii) *or a homogeneous space for an algebraic group G defined over k_1.*

PROOF. We show that in each case the condition (****) holds. In the case i), let $\overline{V}$ be the closure of V, and let $F = \overline{V} - V$. The set F is defined over an algebraic extension K of k_1. An elementary argument then shows that, for every sufficiently large n, there exists a homogeneous polynomial Φ of degree n with coefficients in K which vanishes on F without vanishing on any of the points of the given finite set S. The set U of points of V where Φ does not vanish is then an affine open answering the question (cf. FAC, no. 52).

In case ii), we choose an affine open U_1 of V defined over k_1 and, for every $s \in S$, denote by A_s the set of $g \in G$ such that $g.s \in U_1$. The A_s are non-empty opens of G and thus have a point g in common that one can suppose to be algebraic over k_1. The set $U = g^{-1}U_1$ then answers the question.   □

Remark. The preceding results are entirely analogous to those of chapter III, no. 12, relative to quotients of a variety by a finite group of automorphisms. The method that we have followed moreover amounts to considering V as a variety over k (necessarily reducible—its components correspond to the V^σ) and passing to the quotient by the group of automorphisms defined by the operations h_σ.

21. Principal homogeneous spaces

Let G be an algebraic group defined over a field k. A *homogeneous space* H for G is a non-empty algebraic variety on which the group G acts transitively; in other words, one is given a map $(g, h) \to g.h$ from $G \times H$ to H which is everywhere regular and satisfies the usual identities

$$1.h = h, \qquad g.(g'.h) = (g.g').h,$$

and such that, for every $h \in H$, the map $g \to g.h$ is a surjection from G to H. If H and the map $G \times H \to H$ are defined over k, one says that the homogeneous space H is defined over k or that H is a k-homogeneous space.

[Note that a homogeneous space *is not necessarily* of the form G/G', where G' is an algebraic subgroup of G. Indeed, to identify H with G/G' one must first choose a point $h \in H$ which is rational over k, and such a point may very well not exist. Even if it does exist, we get a map $G/G' \to H$ which in general is not an isomorphism, but only a purely inseparable map.]

A homogeneous space H is called *principal* if $g.h = h$ implies $g = 1$, and if the map which, to every pair (h, h') of points of H, assigns the unique element $g \in G$ such that $h' = g.h$ is a regular map of $H \times H$ to G. If this map is defined over k, one says that H is a principal homogeneous space defined over k.

When H has a point rational over k, say h_0, the map $g \to g.h_0$ is a biregular isomorphism of G to H; one can say that H is the "affine space" associated to the group G. Over an algebraically closed field, there is thus no essential difference between "principal homogeneous space" and "group". It is not the same over an arbitrary field; the classes of principal homogeneous spaces over G form a set analogous to the "Brauer group" and depend on the arithmetic properties of k. More precisely, this set is isomorphic to $H^1(\mathfrak{g}_s, G_s)$ where $\mathfrak{g}_s$ denotes the Galois group of k_s/k (k_s being the separable closure of k), and where G_s denotes the group of points of G rational over K_s. Of course, the cohomology should be defined by *continuous* cochains where one endows $\mathfrak{g}_s$ with its natural topology as a Galois group and G_s with the discrete topology (cf. Lang-Tate [54]). When k is a *finite field*, $H^1(\mathfrak{g}_s, G_s)$ is trivial, as we will see in chap. VI. This is no longer true in the case of a number field or of p-adic fields, cf. for example Tate [84].

22. Construction of the Jacobians $J_{\mathfrak{m}}$ over a perfect field

Let k be a perfect field and let $\overline{k}$ be its algebraic closure. Let X be a curve defined over k and let $\mathfrak{m}$ be a modulus on X; we will suppose that $\mathfrak{m}$ is *rational* over k (cf. §1). Since k is perfect, this means only that the points of the support S of $\mathfrak{m}$ are algebraic over k and that $\mathfrak{m}^{\sigma} = \mathfrak{m}$ for every element σ belonging to the Galois group of $\overline{k}/k$. We are going to show that, with these conditions, the Jacobian $J_{\mathfrak{m}}$ can be defined over k.

We know in any case that $J_{\mathfrak{m}}$ can be defined over the field $\overline{k}$, as can the canonical map $\varphi_{\mathfrak{m}} : X \to J_{\mathfrak{m}}$. As the construction of an algebraic variety requires only a finite number of constants, we immediately deduce the existence of a finite extension k_1/k such that the variety $J_{\mathfrak{m}}$, its group law, and the map $\varphi_{\mathfrak{m}}$ are defined over k_1. After enlarging k_1, we can suppose that k_1 is Galois over k; let $\mathfrak{g}$ be its Galois group.

We are now going to apply the procedure of descent of the base field of no. 20 to the variety $J_{\mathfrak{m}}$. For that, let $\sigma \in \mathfrak{g}$ and let $\theta_{\mathfrak{m}}^{\sigma} : X \to J_{\mathfrak{m}}^{\sigma}$. The map $\varphi_{\mathfrak{m}}^{\sigma}$ admits the modulus $\mathfrak{m}^{\sigma} = \mathfrak{m}$, thus factors as $\varphi_{\mathfrak{m}}^{\sigma} = h_{\sigma} \circ \varphi$, where $h_{\sigma} : J_{\mathfrak{m}} \to J_{\mathfrak{m}}^{\sigma}$ is an "affine" homomorphism. *A priori* the map h_{σ}

is only defined over $\overline{k}$. In fact, if α is a k_1-automorphism of $\overline{k}$, $\varphi_{\mathfrak{m}}^{\alpha} = \varphi_{\mathfrak{m}}$, $\varphi_{\mathfrak{m}}^{\sigma\alpha} = \varphi_{\mathfrak{m}}^{\sigma}$, whence $\varphi_{\mathfrak{m}}^{\sigma} = h_{\alpha}^{\sigma} \circ \varphi_{\mathfrak{m}}$ and the uniqueness of h_{σ} shows that $h_{\sigma}^{\alpha} = h_{\sigma}$, that is to say that h_{σ} is defined over k_1. The same uniqueness shows that the formula $h_{\sigma\tau} = (h_{\sigma})^{\tau} \circ h_{\tau}$ holds. Thus we can effect a descent of the base field by means of the h_{σ} and we get a k-variety that we will designate $J_{\mathfrak{m}}^{(1)}$. We can proceed in the same manner with the homogeneous part h_{σ}^{0} of the affine homomorphisms h_{σ}. Thus we get by descent of the base field another k-variety, which we will denote $J_{\mathfrak{m}}^{(0)}$. From the fact that the h_{σ}^{0} are homomorphisms for the group structure, the group law of $J_{\mathfrak{m}}^{(0)}$ deduced from that of $J_{\mathfrak{m}}$ is *defined over k*. Similarly, we see that $J_{\mathfrak{m}}^{(1)}$ is a *principal homogeneous space defined over k* and that the map $\varphi_{\mathfrak{m}} : X \to J_{\mathfrak{m}}^{(1)}$ is *defined over k*.

Here again, we can characterize $\varphi_{\mathfrak{m}} : X \to J_{\mathfrak{m}}^{(1)}$ by a *universal property*. Generally, let H be a principal homogeneous space for a group G and let f be a map from a curve X to H. If D is a divisor of *degree 0* on X, prime to the set of points where f is not defined, we can define $f(D)$ as an element *of the group G*. In particular, it makes sense to say that f *admits a modulus* $\mathfrak{m}$: we should have $f(D) = 0$ each time that $D \sim_{\mathfrak{m}} 0$.

Proposition 13. *Let $f : X \to H$ be a rational map from a curve X to a principal homogeneous space H for a group G. If f admits the modulus $\mathfrak{m}$, f can be factored as*

$$f = \theta \circ \varphi_{\mathfrak{m}}$$

where $\theta : J_{\mathfrak{m}}^{(1)} \to H$ is an affine homomorphism. This decomposition is unique. Furthermore, if f is defined over an extension k' of k, the same is true of θ.

PROOF. The existence and uniqueness of θ do not involve the base field, and have already been proven (thm. 2). Thus suppose that f is defined over an extension k' of k and let $k'' = k'.\overline{k}$ be the compositum of k' and $\overline{k}$. We first show that θ is defined over k''. Since k'' contains $\overline{k}$, we can identify $J_{\mathfrak{m}}^{(1)}$ with $J_{\mathfrak{m}}$ and the construction of θ given in the proof of thm. 2 (as the composition $J_{\mathfrak{m}} \to X^{(\pi)} \to H$) shows that θ is defined over k''. As k'' is Galois over k', it suffices now to see that $\theta^{\alpha} = \theta$ for every k'-automorphism α of the universal domain. But $f = \theta^{\alpha} \circ \varphi_{\mathfrak{m}}$ since f and $\varphi_{\mathfrak{m}}$ are defined over k', whence $\theta^{\alpha} = \theta$ applying the uniqueness of θ. $\qquad\qquad\square$

Corollary. *The map $\varphi_{\mathfrak{m}} : X \to J_{\mathfrak{m}}^{(1)}$ is characterized up to k-isomorphism by the property of prop. 13.*

Remark. The preceding arguments have a more general application: they show that every "canonical" construction can be effected over the base field of the initial variety when this field is perfect. For example, we return to the situation of no. 5 and let Y be a birational group defined over a perfect field k. Lemma 8 (proved directly by Rosenlicht for an algebraically closed field) shows that Y is birationally isomorphic over $\bar{k}$ to a uniquely defined algebraic group G. Applying descent of the base field to G, we deduce that G can be defined over k, which proves lemma 8 for the case of a perfect field.

23. Case of an arbitrary base field

As the case of a perfect base field will suffice for what follows, we limit ourselves to some brief indications.

Let k be an arbitrary field and let k_s be its separable closure (that is to say the composite of all the separable algebraic extensions of k). Let X be a curve defined over k and let $\mathfrak{m}$ be a modulus rational over k. One begins by constructing the Jacobian $J_\mathfrak{m}$ *over the field* k_s, imitating the construction given in §1 in the case of an algebraically closed field. The construction rests solely on the possibility of choosing a point P_0 away from S and rational over $\bar{k}$, which is still possible over k_s by virtue of the following lemma:

Lemma 26. *Every algebraic variety Y defined over k_s has a point rational over k_s.*

(Applying this result to the opens of Y, we see that these points are *dense* in Y.)

PROOF. We sketch the proof of this well known result. Let K be the field of rational functions of Y over k_s. It suffices to show that, for a model Y' of K, the points of Y' rational over k_s are dense. Let $f_1, \ldots, f_n$ be a separating transcendence basis of K over k_s and let g be a generator of $K/k_s(f_1, \ldots, f_n)$. The function g satisfies an algebraic equation

$$a_0 g^m + \cdots + a_m = 0 \quad a_i \in k_s(f_1, \ldots, f_n),$$

whose derivative does not vanish identically (by the definition of a separating transcendence basis). If we take the subvariety of affine space of dimension $n+1$ defined by the preceeding equation for the model Y' of K, every point of Y' whose first n coordinates belong to k_s and such that neither the derivative of the equation nor a_0 vanish, belongs to k_s. As these points are visibly dense in Y', this proves the lemma. □

Once the Jacobians $J_\mathfrak{m}$ are defined over k_s, descent of the base field from k_s to k is made by exactly the same procedure as in the preceding no. and one can thus prove prop. 13 for an arbitrary field.

Bibliographic note

The theory of the usual Jacobian has its source in the theorems of Abel and Jacobi; from this point of view one could say that the "generalized Jacobi problem" treated by Clebsch and Gordan ([20], §43; see also the exposé of Krazer and Wirtinger [47], §XIII) is at the origin of generalized Jacobians. These are mentioned explicitly for the first time by Severi ([81], chap. II) in the case of ordinary singularities (the base field being C, of course). Severi studies their analytic and algebraic structures (without always separating the two), and observes that these varieties are birationally products of the usual Jacobian with linear varieties. The paper of Rosenlicht [64] takes up the question in the most general case (from the point of view of singularities as well as of base field); most of the results of this chapter are due to him.

The method used by Chow [18] for constructing the usual Jacobian can also be used for generalized Jacobians following Igusa [38]. In the memoir of Igusa, generalized Jacobians appear as "limits" of usual Jacobians; the same is true of the analytic fibrations with exceptional fibers of Kodaira [44] and their algebraic analogues (Néron [119], Raynaud [121]).

The structure of the group of invertible elements of $k[[T]]$ has been clarified by Artin-Hasse, Whaples, Dieudonné, etc. There is a bibliography in the note of Dieudonné [24].

The theorem on descent of the base field was implicitly used in several memoirs of Châtalet (see in particular [14]), but without sufficient justification. It was made explicit and proved by Weil [95].

CHAPTER VI

Class Field Theory

§1. The isogeny $x \to x^q - x$

1. Algebraic varieties defined over a finite field

Let k be a finite field with $q = p^n$ elements and let V be an algebraic variety defined over k (or, as one also says, a k-*variety*). Suppose that V is defined by charts U_i (isomorphic to affine k-varieties) and changes of coordinates u_{ij} (with coefficients in k). If $x = (x_1, \dots, x_r)$ is a point of an affine space, we write Fx, or x^q, for the point with coordinates $(x_1^q, \dots, x_r^q)$. The map $x \to Fx$ commutes with polynomial maps with coefficients in k. In particular, it maps each of the U_i into itself and commutes with the u_{ij}; therefore by "glueing" it operates on V. The image of a point $x \in V$ will again be denoted Fx or x^q.

Let us give another interpretation of this map F:

Generally, let V be an algebraic variety defined over an algebraically closed field K (for example $\bar{k}$, or even a universal domain Ω). The automorphism $x \to x^p$ of K transforms V into a variety that one can denote V^p (at least when there is no confusion with the product of p copies of V) and there is a canonical map $\theta : V \to V^p$. This map is bijective, bicontinuous and identifies the regular functions on V^p with the p-th powers of regular functions on V (which gives a particularly simple description of V^p from the point of view of sheaves). One can say that $\theta : V \to V^p$ is the "maximal height 1 purely inseparable covering" of V^p. (We have already met it several times in this form, cf. chap. III, nos. 8 and 14.)

Repeating this construction n times, we get a map $\theta^n : V \to V^q$ which is purely inseparable of degree $q^{dim.\ V}$. If in addition V is defined over k, the varieties V and V^q are canonically isomorphic, and composing θ^n with this isomorphism we get the map F defined directly above.

Proposition 1. *The k-variety structure of V is unambiguously defined by its structure of variety (over Ω) and by the map F.*

PROOF. Indeed, it is immediate that the rational functions defined over k are characterized by the equation

$$f \circ F = f^q. \qquad \square$$

[Although trivial, this proposition will play an essential role in what follows. Because of it, varieties defined over a finite field are often as easy to study as those defined over an algebraically closed field.]

If W is a subvariety of V defined over an extension K/k, the image FW of W by F is a subvariety of V and W is rational over k if and only if $FW = W$; the same holds for an arbitrary *cycle*. In particular, the set V_k of point of V rational over k is the set of *fixed points* of F.

If V and V' are two varieties defined over k, and if $\varphi : V \to V'$ is a rational map with graph W, there is one and only one rational map, denoted φ^F, with graph FW; φ^F can also be characterized by the formula

$$\varphi^F \circ F = F \circ \varphi.$$

We have $\varphi \circ F = F \circ \varphi \iff \varphi^F = \varphi \iff \varphi$ is defined over k.

All of these properties can be checked immediately starting from one or the other of the definitions of F given above.

2. Extension and descent of the base field

The notations being those of the previous no., let k_1/k be a finite extension of k of degree m. The field k_1 is a field with q^m elements and a map $F_1 : V \to V$ corresponds to it. It is clear that $F_1 = F^m$.

Conversely, given a variety V defined over k_1 and a map $F : V \to V$ of V to itself, we ask under what conditions one can *descend* the base field of V to k (in the sense of chap. V, no. 20) such that F is the corresponding map $x \to x^q$. We will suppose for simplicity that the condition of cor. 1 of prop. 12 of chap. V is satisfied (every finite subset of V which is algebraic over k_1 is contained in an affine open subset of V algebraic over k_1).

Proposition 2. *Under the preceding hypotheses, it is possible to descend the base field if and only if F is of the form $\varphi \circ \theta$, where $\theta : V \to V^q$ is the* *canonical map of V to V^q and where φ is a biregular isomorphism from V^q to V.*

PROOF. The necessity is obvious. To prove the sufficiency, we are going to apply the result of chap. V, *loc. cit.* The Galois group $\mathfrak{g}$ of k_1/k is cyclic of order m, generated by the automorphism defined by $\sigma(\lambda) = \lambda^q$. Descent of the base field is determined by isomorphisms $h_\alpha : V \to V^\alpha$, $\alpha \in \mathfrak{g}$, satisfying the "cocycle" condition

$$h_{\alpha\beta} = (h_\alpha)^\beta \circ h_\beta, \qquad \text{cf. chap. V, no. 20.} \qquad (**)$$

Because the group $\mathfrak{g}$ is cyclic, the system of the h_α is determined when one knows $h = (h_\sigma)^{-1}$, and condition $(**)$ is then

$$h \circ h^\sigma \circ \ldots \circ h^{\sigma^{m-1}} = 1 \qquad \text{(identifying V and $V^{\sigma^m} = V^{q^m}$)}. \qquad (**')$$

We take for h the map φ such that $F = \varphi \circ \theta$. The relation $F^m = F_1$ shows that F commutes with F_1, thus is defined over k_1, and the same is true of $h = \varphi$. It remains to check $(**')$. For this, we first note that, for every map ρ, one has $\rho^\sigma \circ \theta = \theta \circ \rho$. Using this formula and the definition of h, we arrive at

$$\left(h \circ h^\sigma \circ \ldots \circ h^{\sigma^{m-1}}\right) \circ \theta^m = F^m = F_1$$

and since $\theta^m = F_1$ (taking into account the identification $V = V^{q^m}$), this establishes $(**')$. Thus we get a structure of k-variety on V and its very construction shows that the corresponding map F is the given map. $\qquad \square$

3. Tori over a finite field

We indicate, by way of example, how prop. 2 can be applied to the classification of *tori over a finite field k* (such groups are encountered in particular in the local part of generalized Jacobians of curves defined over k).

Thus let V be a torus of dimension r, in other words $(\mathbf{G}_m)^r$. We seek to descend its base field from k_1 to k, in such a way as to obtain an *algebraic group* over k. This amounts to restricting to operations F which are endomorphisms of $(\mathbf{G}_m)^r$; but such an endomorphism corresponds to a square matrix of degree r, with coefficients in $\mathbf{Z}$. Identifying F with the corresponding matrix, the factorization condition of prop. 2 translates to the relation $F = q.\Phi$ where Φ is an invertible matrix (i.e., an element of $GL(r, \mathbf{Z})$); the condition $F^m = F_1$ translates to $\Phi^m = 1$. Thus, k-groups isomorphic over $\overline{k}$ to $(\mathbf{G}_m)^r$ correspond to *elements of finite order* in the group $GL(r, \mathbf{Z})$; two such groups are k-isomorphic if and only if the corresponding matrices are conjugate in $GL(r, \mathbf{Z})$. Thus we get a bijective correspondence between classes of such groups and *classes of representations* of degree r with integral coefficients of a finite cyclic group. When the order of this group is a prime number l, these representations can be completely determined using the ideal class group of the cyclotomic field $\mathbf{Q}(\sqrt[l]{1})$, cf. Reiner [62].

(We point out, following Tate, a more intrinsic definition of the representation associated to a k-group G of the type above: one considers the discrete group $X(G)$ of rational *characters* of G, defined over $\overline{k}$ (séminaire Chevalley [17], exposé 4), and lets the Galois group of $\overline{k}/k$ act on $X(G)$.)

Knowledge of the matrix Φ associated to G permits us to treat various questions about G. For example, the number of points of G rational over k_n, the extension of k of degree n, is

$$\nu_n(G) = \det(q^n - \Phi^n) = \sum_{h=0}^{h=r}(-1)^h q^{n(r-h)} \sum_{i_1<\cdots<i_h}(\lambda_{i_1}\ldots\lambda_{i_h})^n$$

where $\lambda_1,\ldots,\lambda_r$ are the eigenvalues of Φ (which are roots of unity). From this we deduce the computation of the *zeta function* of G:

$$\zeta_G(s;k) = \prod_{h=0}^{h=r}\prod_{i_1<\cdots<i_h}(1 - \lambda_{i_1}\cdots\lambda_{i_h}q^{r-h}t)^{(-1)^{h+1}}$$

where we have put, as usual, $t = q^{-s}$. If we denote the h-th exterior power of the matrix Φ by Φ_h, this formula can be written simply as

$$\zeta_G(s;k) = \prod_{h=0}^{h=r}\det(1 - q^{r-h}t\Phi_h)^{(-1)^{h+1}}. \tag{1}$$

The factors $\det(1 - q^{r-h}t\Phi_h)$ are of the same type as those which occur in Artin's non-Abelian L-functions. This can be made more precise as follows:

Given a finite Galois extension L/K of a number field K, let $\mathfrak{g}$ be its Galois group and let M be a homomorphism of $\mathfrak{g}$ to $GL(r, \mathbf{Z})$. Using the representation M, we descend the base field of the torus $(\mathbf{G}_m)^r$ from L to K, thus obtaining an algebraic group G defined over K. If $\mathfrak{p}$ is a prime ideal of K, the group G defines by reduction modulo $\mathfrak{p}$ a group $G_\mathfrak{p}$ of the type above (this holds for almost all $\mathfrak{p}$). The matrix Φ associated to $G_\mathfrak{p}$ is none other than $M(\sigma_\mathfrak{p})$, where $\sigma_\mathfrak{p} \in \mathfrak{g}$ denotes the Frobenius substitution $(\mathfrak{p}, L/K)$ attached to $\mathfrak{p}$ (defined up to an interior automorphism). Put

$$\zeta_G(s) = \prod_\mathfrak{p}\zeta_{G_\mathfrak{p}}(s) \qquad \text{(defined up to a finite number of factors)};$$

this is the Hasse-Weil zeta function of G. Formula (1) above gives this zeta function explicitly as

$$\zeta_G(s) = \prod_{h=0}^{h=r}L_h(s - r + h)^{(-1)^h} \qquad \text{(up to an elementary factor)} \tag{2}$$

denoting by L_h the Artin L-function attached to the representation of $\mathfrak{g}$ given by the h-th exterior power of the representation M. Note that this function depends only on the representation M from the rational point of view, in other words it does not change when G is modified by an isogeny.

Formula (2) gives a typical example of what happens to a zeta function after a descent of the base field. For other examples (elliptic curves or Abelian varieties with complex multiplication, cubic surfaces), see Deuring [21], Shimura-Taniyama [82], and Weil [92].

4. The map $x \to x^{-1} F x$

Let k be a finite field with q elements, and let G be an algebraic group defined over k. To avoid any confusion between x^q and the q-th power of x in G, we systematically use the notation Fx.

Proposition 3. *The map $x \to x^{-1} F x$ is surjective if G is connected.*

PROOF. More generally, for $y \in G$, consider the map u_y from G to G defined by

$$u_y(x) = x^{-1} y F x.$$

Since $F : G \to G$ factors as $G \to G^q \to G$, its differential is identically zero; from this we deduce that $d(u_y) = d(x^{-1}) y F x$, thus that the tangent map to u_y is *everywhere surjective*. A fortiori, u_y is generically surjective, and $u_y(G)$ contains a non-empty open subset U_y. If now z is any point of G, the hypothesis that G is connected implies $U_z \cap U_e \neq \emptyset$. Let $t \in U_z \cap U_e$. We have $t = x^{-1} z F x$ and $t = y^{-1} F y$, with $x, y \in G$. Putting $u = yx^{-1}$, we find

$$z = u^{-1} F u$$

which proves the proposition. .
$\square$

Corollary 1. *Every homogeneous space for G defined over k has a rational point over k.*

PROOF. Let H be the homogeneous space in question and let $h \in H$. Since H is homogeneous, there exists $g \in G$ such that $h = g.Fh$. By prop. 3, there exists $x \in G$ such that $g = x^{-1}.Fx$; thus $x.h = Fx.Fh$. Because $G \times H \to H$ is defined over k, we have $FxFh = F(xh)$ and the equation $x.h = F(x.h)$ shows that $x.h$ is a point of H rational over k.
$\square$

Examples. Every Severi-Brauer variety [14] over a finite field is trivial. Every linear algebraic group defined over k has a Borel subgroup defined over k (for the set of these subgroups is naturally endowed with a homogeneous space structure).

Corollary 2. *Let* $0 \to G \xrightarrow{u} G' \xrightarrow{v} G'' \to 0$ *be an exact sequence of connected algebraic groups, defined (as well as the homomorphisms u and v) over a finite field k. The sequence*

$$0 \to G_k \xrightarrow{u} G'_k \xrightarrow{v} G''_k \to 0$$

is then exact.

(Recall that, if V is a variety defined over k, we denote by V_k the set of points of V rational over k.)

PROOF. The only non-trivial fact is that G'_k maps *onto* G''_k. Thus for $x'' \in G''_k$, let H be its inverse image in G'; this is a class modulo G. Thus it is a *homogeneous space* for G and the fact that $Fx'' = x''$ shows that $FH = H$, i.e., that H is defined over k. Corollary 1 then shows that H contains a point x' rational over k, which proves the result. $\square$

[Of course, the hypothesis that G is connected is essential. We will see later what happens when G is a finite group.]

5. Quadratic forms over a finite field

We are going to see how prop. 3, applied to the special orthogonal group, gives the classification of quadratic forms over a finite field.

Let V be a vector space of dimension n over k, and let Q be a non-degenerate quadratic form on V. We suppose to start that the characteristic p of k is $\neq 2$. The discriminant Δ of Q is then an element of k^*/k^{*2}, a group which can be identified with $\{+1, -1\}$ by the map $\lambda \to \lambda^{(q-1)/2}$.

Proposition 4. *Two quadratic forms Q and Q' are equivalent if and only if they have the same discriminant.*

(Thus we get a complete classification of quadratic forms over k: such a form is always equivalent to a form of the type $x_1^2 + \cdots + x_{n-1}^2 + g x_n^2$, where g is either a square or a non-square.)

PROOF. We suppose at first that the discriminant Δ is equal to 1, and show that Q is equivalent to the form $x_1^2 + \cdots + x_n^2$ on the vector space $E = k^n$. This equivalence holds in any case over the algebraic closure $\overline{k}$ of k, by the elementary theory of quadratic forms; thus let $u : E \to V$ be an isomorphism from E to V defined over $\overline{k}$. Choosing an orthogonal basis e_i of V, we immediately see that $\det(u)^2 = \Delta$. Then we form the linear map $v = u^{-1} F u$ and we have

$$\det(v) = \Delta^{-1/2} \Delta^{q/2} = \Delta^{(q-1)/2} = +1.$$

Furthermore, it is clear that $v \in O(E)$, the orthogonal group of E. As the special orthogonal group $SO(E)$ is *connected*, prop. 3 applies and permits us to write v as $v = w^{-1}.Fw$, with $w \in SO(E)$. Putting $u' = u.w^{-1}$, we get an isomorphism from E to V which is invariant under F, and is thus defined over k, which proves the proposition in this case.

When the discriminant Δ is not a square, one argues similarly, replacing the form $x_1^2 + \cdots + x_n^2$ by the form $x_1^2 + \cdots + x_{n-1}^2 + gx_n^2$ where g is not a square. □

[Instead of using prop. 3, we could have used cor. 1, remarking that the set of quadratic forms on V with a given discriminant naturally form a *homogeneous space* for the group $SO(E)$.]

When the characteristic is 2, one must separate the cases n even and n odd. In the first case, $O(E)$ has a connected subgroup $O_+(E)$ of index 2 and one finds again two types of quadratic forms, characterized by their *Arf invariant*. In the second case, $O(E)$ is connected and there is only one type of quadratic form. For more details, see Dieudonné [22], chap. I, §16 and chap. II, §10.

6. The isogeny $x \to x^q - x$: commutative case

Suppose now that G is a *commutative* group; we will write the law of composition additively. There is thus no confusion possible in writing x^q in place of Fx.

Proposition 5. *Let k_1 be a finite extension of k and let $\mathfrak{g}$ be the Galois group of k_1/k. If G is a connected, commutative, algebraic group defined over k, then*

$$H^m(\mathfrak{g}, G_{k_1}) = 0 \qquad \text{for all } m \geq 1.$$

(Here we mean the cohomology of the finite group $\mathfrak{g}$ acting in the obvious way on the points of G rational over k_1.)

PROOF. The group $\mathfrak{g}$ being cyclic, the groups $H^m(\mathfrak{g}, G_{k_1})$ depend only on the parity of m. For $m = 1$, we must show that every element of trace zero is of the form $x - x^\sigma$. Thus let $g \in G_{k_1}$ so that

$$g + g^q + \cdots + g^{q^{n-1}} = 0, \qquad \text{with } n = [k_1 : k].$$

By prop. 3, we can write $g = x^q - x$, with $x \in G$. The preceding formula then gives

$$(x^q - x) + \cdots + (x^{q^n} - x^{q^{n-1}}) = 0$$

so $x^{q^n} = x$; this means that $x \in G_{k_1}$, and $H^1(\mathfrak{g}, G_{k_1}) = 0$. But because G_{k_1} is a finite group, *Herbrand's lemma* ([16], §10) shows that $H^2(\mathfrak{g}, G_{k_1})$ has the same order as $H^1(\mathfrak{g}, G_{k_1})$, which proves the proposition. □

Corollary. *Every element of G_k is the trace of an element of G_{k_1}.*

PROOF. This merely expresses the fact that $H^2(\mathfrak{g}, G_{k_1})$ is trivial.   □

DIRECT PROOF. Let $\varphi(x) = x + x^q + \cdots + x^{q^{n-1}}$. The differential of the homomorphism φ is equal to that of x, which shows that φ is surjective. If $y \in G_k$, we choose $x \in G$ such that $\varphi(x) = y$ and the equation $y^q = y$ shows that $x \in G_{k_1}$.   □

We return to the map $\wp(x) = x^q - x$. The proof of prop. 3 (based on the computation of the differential of $\wp$) shows that $\wp$ is *separable*. Furthermore, since G is commutative, it is a homomorphism of G to itself. Its kernel is the set of $x \in G$ such that $x^q = x$, which is G_k. Thus, we have an exact sequence

$$0 \to G_k \to G \overset{\wp}{\to} G \to 0.$$

This exact sequence makes G a *covering* of itself which is clearly Abelian (over k), the Galois group being the group of translations $x \to x + a$, $a \in G_k$. We are going to see that this covering is the *largest* which enjoys these properties (which suggests an analogy with the "absolute class field" of Hilbert—the precise relation between them will be discussed in §§4,5, and 6).

Proposition 6. *Let $\theta : G' \to G$ be a separable isogeny defined over k. The following four conditions are equivalent:*

i)   *The extension $k(G')/k(G)$ defined by θ is Galois.*
ii)   *The extension $k(G')/k(G)$ defined by θ is Abelian.*
iii) *The kernel of θ is contained in G'_k.*
iv)   *The isogeny θ is a quotient of the isogeny $\wp : G \to G$.*

If these conditions are satisfied then the Galois group of the extension $k(G')/k(G)$ is the group of translations $x \to x + a$, where a runs through the kernel of θ.

PROOF. ii) $\Longrightarrow$ i) trivially. i) $\Longrightarrow$ iii) because, if σ_i are the elements of the Galois group, the σ_i transform every rational point into a rational point and as the elements of $\theta^{-1}(0)$ are the transforms of 0, they are rational. Finally, iii) $\Longrightarrow$ ii) because the translations $x \to x + a$, $a \in \theta^{-1}(0)$, are defined over k and are $k(G)$-automorphisms of $k(G')$, equal in number to the degree of the extension. Thus i), ii) and iii) are equivalent.

We have iv) $\Longrightarrow$ ii) because every subextension of an Abelian extension is Abelian. Conversely, suppose iii) is true and let $\wp' : G' \to G'$ be the isogeny $x \to x^q - x$ of G'. Because θ is defined over k, there is a commutative diagram

$$
\begin{array}{ccc}
G' & \xrightarrow{\ \wp'\ } & G' \\
\theta \downarrow & & \theta \downarrow \\
G & \xrightarrow{\ \wp\ } & G.
\end{array}
$$

As the kernel of θ is contained in that of $\wp'$, namely G'_k, the map $\wp'$ defines by passage to the quotient a homomorphism $\alpha : G \to G'$, and thus we have factored $\wp : G \to G$ as $G \xrightarrow{\alpha} G' \xrightarrow{\theta} G$. $\qquad\qquad\square$

Remark. When the group G is no longer assumed to be commutative, the map

$$
\wp(x) = x^{-1} F x
$$

is not in general a homomorphism. In any case, $\wp(x) = \wp(y)$ if and only if $x \equiv y \bmod G_k$ and $\wp$ defines an isomorphism from the *homogeneous space* G/G_k to G. In particular, $\wp : G \to G$ is an *unramified Galois covering*, which has properties very close to those of an isogeny.

§2. Coverings and isogenies

7. Review of definitions about isogenies

Let k be a field, $\overline{k}$ its algebraic closure and let V be a normal, irreducible, algebraic variety defined over $\overline{k}$. Let $K = \overline{k}(V)$ be the field of rational functions on V. One knows that the notion of a *covering* of V is a *birational* notion, in other words it depends only on the field K. More precisely, let L/K be a finite separable extension of K (we limit ourselves to separable coverings, because we want to study those which are *Abelian*). The *normalization* W of V in L is the variety whose local rings are those obtained by decomposing the integral closure in L of the local rings $\mathcal{O}_P$ of points $P \in V$. One can similarly define *reducible coverings*, taking instead of the field L a product $\prod L_i$ of separable extension fields L_i of K (in other words, one is given a *separable commutative algebra* over K).

The variety W thus defined comes with a projection $\pi : W \to V$. One says that it is the *covering of V corresponding to the extension L/K*; we will carry over to W all the terminology of extensions of fields: one says that the covering W is Galois, Abelian, of degree n, etc., if L/K is so.

If W is Galois with Galois group $\mathfrak{g}$, the elements $\sigma \in \mathfrak{g}$ define automorphisms of W and V is identified with the quotient variety $W/\mathfrak{g}$ (cf. chap.

III, no. 12). Conversely, if $\mathfrak{g}$ is a group of automorphisms of a normal variety W, the projection $W \to W/\mathfrak{g}$ makes W a Galois covering of $W/\mathfrak{g}$ with Galois group $\mathfrak{g}$.

One says that a point $P \in V$ is *unramified* in W if it is the image under π of n points of W (with $n =$ degree of W). If no point P is ramified, one says that the covering is *everywhere unramified*. If the covering is Galois with Galois group $\mathfrak{g}$, this is equivalent to saying that the elements of $\mathfrak{g}$ other than the identity element act on W without fixed points.

Let V' be another normal variety and let $f : V' \to V$ be a rational map such that $f(V')$ is not contained in the set of ramification points of V. After removing these points, there inverse images by f, as well as the points where f is not regular, we can suppose that $W \to V$ is everywhere unramified and that f is everywhere regular. Then in the product $W \times V'$, let W' be the set of pairs (v', w) such that $\pi(w) = f(v')$. It is immediate that the canonical projection $\pi' : W' \to V'$ makes W' an unramified covering of V', of the same degree as W, Galois if W is and with the same Galois group. One calls it (as is common in topology) the *pull-back* of W by f, and denotes it $f^*(W)$. Of course, this covering is not necessarily irreducible; it decomposes in general into irreducible components which are conjugate if the covering is Galois. In the particular case where V' is a subvariety of V and where f is the canonical injection of V' into V, the covering W' is none other than $\pi^{-1}(V')$.

All of this is relative to an algebraically closed field $\overline{k}$. If W and V are endowed with k-variety structures and if π is defined over k, one says that the covering is *defined over* k. It corresponds to an extension $k(W)/k(V)$. Note that $W \to V$ may very well be Galois over $\overline{k}$ without being Galois over k (cf. prop. 6).

8. Construction of coverings as pull-backs of isogenies

Let G be an algebraic group, N a finite subgroup, and $G/N = H$ the quotient homogeneous space (we do not assume that G is commutative). The canonical map $\pi : G \to H$ makes G an unramified Galois covering of H with group N. If G is defined over a field k, and if each $n \in N$ is rational over k, this covering is defined over k, and Galois over k. In the particular case where G is commutative, the covering π is an *isogeny* (cf. no. 5).

We propose to construct, for any every finite group N, a group $G(N)$ containing N, defined over the prime field $\mathbf{F}_p$, and playing a *universal* role for all Galois coverings with Galois group N. More precisely, let $A(N)$ be the group *algebra* of the group N (over the universal domain, to fix ideas) and let $G(N)$ be the set of invertible elements of $A(N)$. This is an open set in the affine space $A(N)$ (it is the set of systems $\{a_s\}_{s \in N}$ such that $\det(a_{st}) \neq 0$). It is defined over the prime field $\mathbf{F}_p$ and is clearly an irreducible algebraic group containing N. We have:

Proposition 7. *If* $\pi : W \to V$ *is a Galois covering with Galois group* N, *there exists a map* $f : V \to G(N)/N$ *such that* W *is isomorphic to the covering* $f^*(G(N))$.

If further the covering W *is defined and Galois over the field* k, *the map* f *and the isomorphism* $W \to f^*(G(N))$ *can be defined over* k.

PROOF. This is a purely birational question. Thus let K be the field of functions of V, L the field (or rather the algebra, if W is reducible) of functions on W. To find f and an isomorphism $W \to f^*(G)$ comes down to finding a pair of maps $g : W \to G(N)$ and $f : V \to G(N)/N$ such that the diagram

$$
\begin{array}{ccc}
W & \xrightarrow{\ g\ } & G(N) \\
\downarrow & & \downarrow \\
V & \xrightarrow{\ f\ } & G(N)/N
\end{array}
$$

is commutative, and such that g commutes with the action of N. This last condition, together with the structure of $G(N)$, shows that g is of the form $x \to (\varphi^s(x))$ where φ is a rational function on W. The only condition to impose on φ is that g maps W into $G(N)$ [and not just into $A(N)$], that is to say that $\det(\varphi^{st})$ is not identically zero. The existence of such a function then follows from the *normal basis theorem* applied to the Galois extension L/K (one checks immediately that the normal basis theorem extends to Galois *algebras* over a field). As for the map f, it is deduced from g by passage to the quotient.

When the covering is defined and Galois over k, one applies the same argument to the extension $k(W)/k(V)$. $\square$

Corollary. *Every Abelian covering is the pull-back of an isogeny.*

This is the result stated in chap. I, thm. 4.

9. Special cases

The group $G(N)$ is a particular case of the groups defined as sets of invertible elements of a finite dimensional algebra (these are the "bilinear groups" of Elie Cartan). When the algebra has no radical and the ground field is algebraically closed, these groups are products of general linear groups GL_n. Outside this case, they are not well understood. Let us limit ourselves to considering two special cases (which are sufficient for the subsequent applications):

i) N *is cyclic of order* n, *prime to the characteristic* p.

The algebra $A(N)$ is isomorphic to $k[T]/(1 - T^n)$. If k contains a primitive n-th root of unity, call it ϵ, $1 - T^n$ can be decomposed into a product of linear factors and $G(N)$ *is isomorphic to a product of groups* $\mathbf{G}_m$. Thus, for an arbitrary field, $G(N)$ is obtained by *descent of the base field* from a torus $(\mathbf{G}_m)^n$.

Return to the case where $\epsilon \in k$. One of the projections $G(N) \to \mathbf{G}_m$ is given by $T \to \epsilon$. Denoting by θ the isogeny $\mathbf{G}_m \to \mathbf{G}_m$ defined by $\theta(\lambda) = \lambda^n$, we have a commutative diagram

$$
\begin{array}{ccc}
G(N) & \longrightarrow & \mathbf{G}_m \\
\downarrow & & {\scriptstyle\theta}\downarrow \\
G(N)/N & \longrightarrow & \mathbf{G}_m
\end{array}
$$

which shows that the isogeny $G(N) \to G(N)/N$ is the pull-back of the isogeny $\theta : \mathbf{G}_m \to \mathbf{G}_m$. We conclude from this that proposition 7 is valid with θ in place of $G(N) \to G(N)/N$. This is just *Kummer theory*.

When we do not suppose that k contains ϵ, Kummer theory no longer applies. However, we can still, in certain cases, reduce the dimension of $G(N)$. When $n = 3$ for example, we can take as quotient of $G(N)$ the orthogonal group G of the quadratic form $x^2 - xy + y^2$. One sees easily that this group contains a cyclic subgroup N of order 3, formed by the rational points over the prime field, and that the isogeny $G \to G/N$ has the universal property of prop. 7. From the point of view of field theory, this amounts to following statement, which is easy to check directly:

In characteristic different from 3, every cyclic extension of degree 3 can be generated by an element g with conjugates $1/(1 - g)$ and $1 - 1/g$.

ii) N *is cyclic of order* p^n.

Again $A(N) = k[T]/(1 - T^{p^n})$. Changing T to $1 - T$, we see that $A(N)$ is isomorphic to the algebra $k[T]/(T^{p^n})$ of *truncated formal series of order* p^n. According to the corollary to prop. 9 of chap. V, $G(N)$ is thus isomorphic to the product of $\mathbf{G}_m$ by Witt groups W_{n_i}. All the n_i are less than n except one which is equal to n. Projecting $G(N)$ on the corresponding group W_n, we deduce as above that the isogeny $\wp : W_n \to W_n$ can replace the isogeny $G(N) \to G(N)/N$ in prop. 7. This is just *Witt-Artin-Schreier theory* [99]. Note that, contrary to what happens in Kummer theory, no hypothesis on k is necessary.

10. Case of an unramified covering

We return to the situation of prop. 7 and let f be the desired map from V to $G(N)/N$. If f is regular at a point $P \in V$, the covering $f^*(G(N))$ is unramified at P and the same is thus true of W which is isomorphic to it. Conversely:

Proposition 8. *If P is a point of V which is unramified in W, the map f of prop. 8 can be choosen to be regular at P.*

(If the covering $\pi : W \to V$ is defined and Galois over a field k, we also require that f and the isomorphism $W \to f^*(G(N))$ be defined over k.)

PROOF. In view of the construction of f given in the proof of prop. 7, everything comes down to showing that we can choose the normal basis φ^s of L/K such that φ is regular at the points $Q \in W$ mapping to P. Let $\mathcal{O}_P$ be the local ring at P in $k(V)$, $\mathfrak{m}_P$ its maximal ideal and $\mathcal{O}'_P$ its integral closure in $k(W)$, i.e., the intersection of the $\mathcal{O}_Q$ for $Q \in W$ mapping to P. Let $k(P) = \mathcal{O}_P/\mathfrak{m}_P$ and let $k'(P) = \mathcal{O}'_P/\mathfrak{m}_P\mathcal{O}'_P$. Because P is unramified, $k'(P)$ is a Galois algebra over $k(P)$, of degree n equal to that of the covering. Let $\{\lambda^s\}_{s \in N}$ be a normal basis of this algebra. We choose a representative φ of λ in $\mathcal{O}'_P$ and let $\Psi = \det(\varphi^{st})$. The image of Ψ in $k'(P)$ is equal to $\det(\lambda^{st})$, which is an invertible element of $k'(P)$. Thus Ψ is invertible in $\mathcal{O}'_P$ and the φ^s form a normal basis having the required properties. $\qquad\square$

Remark. Applied to the case of a cyclic covering, proposition 8 shows that the generators of Kummer or of Witt-Artin-Schreier can be chosen to be regular at P (provided that P is unramified in W of course). This is a well-known result (cf. [**77**], no. 15, for example).

11. Case of curves

Although it is by no means essential, in this no. we assume that the base field k is algebraically closed. The most interesting case left aside is that of a finite field which will be treated in detail in §6.

So let X be a complete, irreducible, non-singular curve defined over k and let K be the field of rational functions on X. If $Y \to X$ is an Abelian covering of X with Galois group N, the corollary to proposition 7 shows that Y is of the form $f^*(G)$, where $G \to H$ is a separable isogeny with kernel N and f is a rational map of X to the group H. By virtue of the results of chap. V, the map f can be factored as $X \to J_\mathfrak{m} \to H$, where $J_\mathfrak{m} \to H$ is a homomorphism of a generalized Jacobian $J_\mathfrak{m}$ of X to the group H (*a priori* this is only true up to a translation; but one sees immediately that a translation does not change an isogeny when the base field is algebraically closed). The pull-back of the isogeny $G \to H$ by the homomorphism $J_\mathfrak{m} \to H$ is an isogeny $J' \to J_\mathfrak{m}$ and the covering $Y \to X$ is the pull-back of J'. We have thus proved:

Proposition 9. *Every Abelian covering of X is the pull-back of a separable isogeny of a generalized Jacobian of X.*

We are going to complete this result by showing that the isogeny in question is essentially *unique*. To this end it is convenient to introduce the group $\mathrm{Ext}(J_\mathfrak{m}, N)$ of *classes of extensions of $J_\mathfrak{m}$ by N* (see the definition of this group in chap. VII, no. 1). An isogeny $J' \to J_m$ with kernel N is represented by an element j' of this group. An analogous construction allows us to define the group $\mathrm{Cov}(X, N)$ of coverings of X with Galois group N; we have $\mathrm{Cov}(X, N) = \mathrm{Hom}(G_K, N)$, where G_K denotes the Galois group of the maximal separable extension of K. The operation $J' \to \varphi^*(J')$ defines a homomorphism φ^* of $\mathrm{Ext}(J_m, N)$ to $\mathrm{Cov}(X, N)$ and the uniqueness result we have in mind can be stated as follows:

Proposition 10. *For every modulus m and every finite Abelian group N, the homomorphism*

$$\varphi^* : \mathrm{Ext}(J_m, N) \to \mathrm{Cov}(X, N)$$

is injective.

PROOF. Since φ^* is a homomorphism, it suffices to show that $\varphi^*(j') = 0$ implies $j' = 0$. Thus, let $J' \to J_m$ be an isogeny whose pull-back Y decomposes into n irreducible components $Y_1, \dots, Y_n$ (n being the number of elements of the finite group N). Each of the coverings $Y_i \to X$ has degree 1, which shows the existence of a "section" $s : X \to Y$. Since $Y = \varphi^*(J')$, the section s defines a rational map $\psi : X \to J'$ lifting the map $\varphi : X \to J_m$. The maps ψ and φ are regular outside the support S of m; since φ admits the modulus m, prop. 14 of chap. III shows that ψ also admits the modulus m, thus factors as $\theta \circ \varphi$, where θ is a homomorphism from J_m to J' (up to a translation). The composition $J_m \overset{\theta}{\to} J' \to J_m$ is the identity (for it is the identity on the image of X, which generates J_m); this shows that J' is isomorphic to the product $J_m \times N$, in other words that the isogeny J' is trivial.   □

We easily deduce from this result that *the number of irreducible components of $\varphi^*(J')$ is equal to that of J'.*

12. Case of curves: conductor

We keep the notations and hypotheses of the preceding no.

Proposition 11. *Let $\pi : Y \to X$ be an Abelian covering of the curve X. Then there exists a smallest modulus m such that Y is the pull-back of an isogeny $J' \to J_m$ and the support of this modulus is equal to the set of points $P \in X$ which ramify in Y.*

(It is this modulus that one calls the *conductor* of the extension L/K corresponding to the covering Y.)

Lemma 1. *Let* $\mathfrak{m}'$ *and* $\mathfrak{m}''$ *be two moduli such that* Y *is the pull-back of an isogeny of* $J_{\mathfrak{m}'}$ *and of an isogeny of* $J_{\mathfrak{m}''}$. *Then* Y *is the pull-back of an isogeny of* $J_{\mathfrak{m}}$ *where* $\mathfrak{m} = \mathrm{Inf}(\mathfrak{m}', \mathfrak{m}'')$.

Let us admit this lemma for a moment. The existence of a smallest modulus $\mathfrak{m}$ such that Y is the pull-back of an isogeny of $J_{\mathfrak{m}}$ is then clear. Let S be the support of $\mathfrak{m}$ and let S' be the set of points $P \in X$ which ramify in Y. Clearly $S' \subset S$. Conversely, if $P \notin S'$, prop. 8 shows that there exists a map $f : X \to H$ of X to a commutative group H, and an isogeny $G \to H$ such that $f^*(G)$ is isomorphic to Y and such that f is regular at P. According to the results of chap. V, f factors as $X \to J_{\mathfrak{m}'} \to H$, where $\mathfrak{m}'$ is a modulus whose support does not contain P. In view of the minimality of $\mathfrak{m}$ we have $\mathfrak{m}' \geq \mathfrak{m}$, which shows that $P \notin S$ and proves the proposition.   □

We pass to the proof of the lemma. Suppose that $\deg(\mathfrak{m}') \geq 1$ and $\deg(\mathfrak{m}'') \geq 1$, otherwise there is nothing to prove. Put $\mathfrak{m}_1 = \mathrm{Sup}(\mathfrak{m}', \mathfrak{m}'')$. If J denotes the usual Jacobian, there is a sequence of canonical homomorphisms

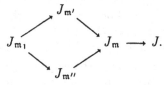

We denote by H', H'', H and H_1 the kernels of the canonical homomorphisms from $J_{\mathfrak{m}_1}$ to $J_{\mathfrak{m}'}$, $J_{\mathfrak{m}''}$, $J_{\mathfrak{m}}$ and J. Then $H' \subset H$, $H'' \subset H$ and $H \subset H_1$. The structure of each of these groups can be determined immediately using the results of chap. V, §3. They are products of groups of the type $U_P^{(n)}/U_P^{(n')}$. We conclude from this that H' and H'' *generate* H, and that the canonical map $H' \times H'' \to H$ *is a biregular isomorphism* (if $\deg(\mathfrak{m}) \geq 1$), or *identifies* H *with a quotient* $H' \times H''/\mathbf{G}_m$ (if $\mathfrak{m} = 0$).

This being the case, let $J' \to J_{\mathfrak{m}'}$ be an isogeny of $J_{\mathfrak{m}'}$ having Y for pull-back, and let $J'' \to J_{\mathfrak{m}''}$ be an isogeny of $J_{\mathfrak{m}''}$ having the same property. These isogenies have pull-backs J_1' and J_1'' over $J_{\mathfrak{m}_1}$; by virtue of proposition 10 (applied to $\mathfrak{m}_1$), J_1' and J_1'' are isomorphic. We denote them both by J_1. The isogeny $J_1 \to J_{\mathfrak{m}_1}$ is clearly trivial on H' and on H''. Let $s' : H' \to J_1$ and $s'' : H'' \to J_1$ be section homomorphisms over these groups; their sum is a homomorphism s from $H' \times H''$ to J_1. Let Q be the kernel of $H' \times H'' \to H$. The commutative diagram:

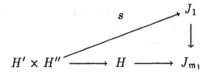

shows that s maps Q to the kernel N of the isogeny $J_1 \to J_{\mathfrak{m}_1}$. Since Q is connected (in fact, we have seen that Q is either 0 or isomorphic to $\mathbf{G}_m$), $s(Q) = 0$, and, since H is identified with $H' \times H''/Q$, the homomorphism s defines, by passage to the quotient, a section homomorphism from H to J_1. Thus the isogeny $J_1 \to J_{\mathfrak{m}_1}$ is trivial on H. It follows easily from this (it is a special case of the Ext exact sequence, cf. chap. VII, prop. 2) that this isogeny is the pull-back of the isogeny $G \to J_{\mathfrak{m}_1}/H = J_{\mathfrak{m}}$, which proves the proposition.   □

Corollary. *Unramified Abelian coverings of an algebraic curve are in one to one correspondence with isogenies of its Jacobian.*

PROOF. This is the special case $\mathfrak{m} = 0$.   □

Examples. 1) Suppose that the Galois group N of the covering $Y \to X$ is of order *prime to p*. The conductor $\mathfrak{m}$ is then equal to the *sum of the points of ramification* each with coefficient 1: this can be seen by applying Kummer theory and prop. 6 of chap. III. (One can also use the structure of generalized Jacobians; it amounts to the same.)

2) Suppose that N is cyclic of order p and let $P \in X$ be a point of ramification of the covering. One can choose an Artin-Schreier generator f which has a pole of order prime to p at P; let n be this order. If c_P denotes the coefficient of P in the conductor, prop. 5 of chap. III shows that $c_P \leq n+1$. Furthermore, one easily checks that the isogeny of $U_P^{(n)}/U_P^{(n+1)}$ defined by f is non-trivial; thus $c_P = n+1$ which determines the conductor, and shows that it coincides with that of Hasse [31]; see also [123], no. 4.4.

§3. Projective system attached to a variety

In all of this §, the letter k denotes a *perfect* field; the algebraic closure of k is denoted by $\overline{k}$.

13. Maximal maps

We make the following conventions:

All varieties considered in this no. and in the following are defined over k and are irreducible. The term *group* is reserved for commutative algebraic groups (irreducible and defined over $\overline{k}$, of course). The term *principal homogeneous space* is reserved for principal homogeneous spaces (in the sense of chap. V, no. 21) for such groups.

If H and H' are two principal homogeneous spaces corresponding to groups G and G', a map $h : H \to H'$ will be said to be a *morphism* if it is an "affine" map, i.e., if

$$h(x + g) = h(x) + h_0(g), \qquad x \in H, g \in G,$$

where $h_0 : G \to G'$ is a homomorphism (of algebraic groups). One says that h is an *isogeny* if h_0 is one. This is the same as saying that H and H' have the same dimension and that h is surjective. The *kernel* of h is the kernel in the usual sense of h_0; it is a subgroup of G.

Definition 1.   Let V be a variety and let $\alpha : V \to H$ be a rational map from V to a principal homogeneous space H. One says that α is maximal if the following condition is satisfied:

(M) — if α factors as $V \xrightarrow{\alpha'} H' \xrightarrow{h} H$, where $\alpha' : V \to H'$ is a rational map and $h : H' \to H$ is a morphism with finite kernel, then h is an isomorphism from H' to H.

Examples. The canonical map from V to its Albanese variety (cf. [52]) is a maximal map (indeed, when h has a finite kernel, H' is an Abelian variety and the fact that h is an isomorphism then follows from the universal property of the Albanese variety). The same is true of the map from a curve to one of its generalized Jacobians (cf. §6).

Factoring h as a surjection and an injection, we see that condition (M) decomposes into two conditions:

(M_1) — $\alpha(V)$ *generates* H (in other words $\alpha(V)$ is not contained in any affine subspace of H distinct from H).

(M_2) — *The map α does not lift to any non-trivial isogeny of H* (in other words, if $\alpha = h \circ \alpha'$, and if h is an isogeny, then h is an isomorphism).

In fact:

Lemma 2.   *Condition (M_2) implies condition (M_1). (It is thus equivalent to the condition (M).)*

PROOF. Suppose that (M_1) is not satisfied, i.e., that there exists an affine subspace H_1 of H, distinct from H and containing $\alpha(V)$. Let G_1 be the subgroup of G corresponding to H_1. If we take H_1 (thus also G_1) maximal, the structure of commutative algebraic groups (chap. III, no. 7) shows that G/G_1 is isomorphic either to the additive group $\mathbf{G}_a$ or to the multiplicative group $\mathbf{G}_m$ or to a simple Abelian variety A. In each of these cases, there exists a non-trivial isogeny $G'' \to G/G_1$; this is clear for $\mathbf{G}_a$ and $\mathbf{G}_m$, and, for A, we can take multiplication by an integer > 1. Let G' be the pullback of this isogeny by $G \to G/G_1$, that is the subgroup of $G \times G''$ formed of the pairs having the same image in G/G_1. Since G' is an extension of G_1 by G'', it is a connected group and the isogeny $G' \to G$ is non-trivial. Choosing an origin in H_1, we can identify H_1 and H with G_1 and G and thus we get an isogeny $h : H' \to H$ corresponding to $G' \to G$. As G_1 embeds in G', the homogeneous space H_1 embeds in H', and the map α factors as $V \to H' \to H$, which contradicts the condition (M_2).   □

Remark. In the preceding proof, the isogeny $h : H' \to H$ can be chosen at will to be separable or purely inseparable of height 1 (in characteristic $p \neq 0$).

Lemma 3. *Every rational map $\alpha : V \to H$, where H is a principal homogeneous space, admits a factorization*

$$V \xrightarrow{\alpha'} H' \xrightarrow{h} H$$

where α' is maximal and where h is a morphism. One can further require h to have a finite kernel.

PROOF. Let H_1 be an affine subvariety of H which is minimal among all those containing $\alpha(V)$. After replacing H by H_1, we can suppose that $\alpha(V)$ *generates* H. If α is not maximal, we can then factor it as

$$V \to H_1 \to H$$

where $H_1 \to H$ is a non-trivial isogeny. Likewise, if $V \to H_1$ is not maximal, we can factor it as $V \to H_2 \to H_1$, etc. Everything comes down to showing that this process stops, in other words that one cannot have a sequence of factorizations

$$V \to H_{n+1} \to H_n$$

where $H_{n+1} \to H_n$ is a non-trivial isogeny.

Denote by G and G_n the groups associated to the homogeneous spaces H and H_n. For every integer $r \geq 1$, we denote by $S_r \alpha_n$ the rational map from V^{2r} to G_n defined by the formula

$$S_r \alpha_n(x_1, y_1, \ldots, x_r, y_r) = \sum_{i=1}^{i=r} \alpha_n(x_i) - \alpha_n(y_i).$$

Since $\alpha(V)$ generates H, there exists an integer r such that $S_r \alpha$ is a *generically surjective* map from V^{2r} to G. Because $G_n \to G$ is an isogeny, the same is true for $S_r \alpha_n$ for all n. The field $\overline{k}(V^{2r})$ of rational functions on V^{2r} then contains a strictly increasing sequence of subfields

$$\overline{k}(G) \subset \overline{k}(G_1) \subset \cdots \subset \overline{k}(G_n) \subset \cdots \subset \overline{k}(V^{2r})$$

which is absurd, since every subfield of a field of finite type is of finite type (cf. [**51**], p. 64).   □

Lemma 4. *Let $\alpha : V \to H$ and $\alpha' : V \to H'$ be two maximal maps. If there exists a morphism $h : H' \to H$ such that $\alpha = h \circ \alpha'$, this morphism is unique, its kernel N is connected, and it defines by passage to the quotient an isomorphism from H'/N to H.*

PROOF. If h_1 and h_2 satisfy $\alpha = h_1 \circ \alpha'$ and $\alpha = h_2 \circ \alpha'$, the set of points where h_1 and h_2 are equal is an affine subvariety of H' containing the image of V, thus equal to H', since α' satisfies (M_1). Now let N_0 be the connected component of the kernel N of h; one can factor h as $H' \to H'/N_0 \to H$, and the morphism $H'/N_0 \to H$ has kernel isomorphic to N/N_0 thus is finite. Since α satisfies (M), it follows that $H'/N_0 \to H$ is an isomorphism, so $N = N_0$ as was to be shown. $\square$

Definition 2. If α and α' are two maximal maps satisfying the condition of lemma 4, one says that α' *dominates* α and one writes $\alpha' \geq \alpha$.

We denote by L the set of maximal maps from V to principal homogeneous spaces. Equipped with the relation $\alpha' \geq \alpha$, the set L is a *preordered* set. If $\alpha' \geq \alpha$ and $\alpha \geq \alpha'$, we write $\alpha' \approx \alpha$; this means that there exists an isomorphism h (necessarily unique, by virtue of lemma 4) from H' to H such that $\alpha = h \circ \alpha'$.

Lemma 5. *The set L is "reticulated" (in other words, every pair of elements of L has a lower bound and an upper bound).*

PROOF. We first show that L is *increasingly filtered*. Let $\alpha_1 : V \to H_1$ and $\alpha_2 : V \to H_2$ be two maximal maps. By virtue of lemma 3, we can factor the map $\alpha_1 \times \alpha_2 : V \to H_1 \times H_2$ by means of a maximal map α and it is clear that $\alpha \geq \alpha_1$ and $\alpha \geq \alpha_2$.

Now we show that, for every $\alpha \in L$, the set $L(\alpha)$ of elements $\beta \leq \alpha$ is *reticulated*. According to lemma 4, the elements of $L(\alpha)$ correspond biuniquely to connected subgroups of the group G associated to the homogeneous space H, and this set is reticulated. To prove that L itself is reticulated, it then suffices to show that, if $\alpha' \geq \alpha$, the operations Sup and Inf of $L(\alpha)$ are induced from those of $L(\alpha')$. This is clear for the operation Inf. Thus let $\beta, \gamma \in L(\alpha)$ and let δ (resp. δ') be their upper bound in $L(\alpha)$ (resp. $L(\alpha')$). Since $L(\alpha) \subset L(\alpha')$, $\delta \geq \delta'$ and we deduce that $\delta' \in L(\alpha)$, whence $\delta' \geq \delta$, as was to be shown. $\square$

14. Some properties of maximal maps

Let $\alpha : V \to H$ be a rational map from the irreducible variety V to a principal homogeneous space H and let G be the group associated to H. Let ω be a differential 1-form on G which is invariant by translations (cf.

chap. III, no. 11). By choosing a point of H, we identify H and G and ω defines a differential form on H which is independent of the point chosen; we again denote this differential form by ω. The inverse image $\alpha^*(\omega)$ of ω by α is a differential form on V.

Proposition 12. *If α is maximal, the relation $\alpha^*(\omega) = 0$ implies $\omega = 0$.*

PROOF. If the characteristic is $\neq 0$, the proposition follows from the fact that α is "inseparably maximal", i.e., does not lift to any purely inseparable isogeny $H' \to H$ (cf. [**78**], thm. 4). If the characteristic is zero, we use the fact that $\alpha(V)$ generates H. We construct a generically surjective map $S_r\alpha : V^{2r} \to G$ (cf. the proof of lemma 3). The differential form $(S_r\alpha)^*(\omega)$ is determined using prop. 17 of chap. III: it is the direct sum of $2r$ terms of the form $\pm\alpha^*(\omega)$ and is thus zero if $\alpha^*(\omega)$ is. As $S_r\alpha$ is generically surjective, it follows that $\omega = 0$.   □

Before stating the following proposition, we observe that, if V is *normal*, and if $\alpha : V \to H$ factors as $V \xrightarrow{\alpha'} H' \xrightarrow{h} H$, where α' is rational and h is an isogeny, then α' is regular if (and only if) α is; this is an immediate consequence of "Zariski's main theorem" (cf. [**51**], chap. V).

Suppose then that V is normal, the map α is regular and also that $k = \mathbf{C}$. We have:

Proposition 13. *In order that α be maximal, it is necessary and sufficient that the homomorphism $\alpha_* : H_1(V) \to H_1(H)$ defined by α be surjective.*

(We have written $H_1(V)$ and $H_1(H)$ for the homology groups of V and H of dimension 1 with coefficients in $\mathbf{Z}$.)

PROOF. The proof is very similar to that of prop. 11 of chap. V, so we limit ourselves to a sketch. We note first that α_* is surjective if and only if there does not exist a subgroup of finite index > 1 of $H_1(H)$ containing $\alpha_*(H_1(V))$, i.e., if α does not factor as $V \xrightarrow{\alpha'} H' \xrightarrow{h} H$, where $\alpha' : V \to H'$ is continuous and where $H' \xrightarrow{h} H$ is a finite covering of degree > 1. According to lemma 24 of chap. V, this covering is in fact an isogeny, and according to lemma 25 of chap. V, the map α' is regular. The proposition follows from this and the fact that if α' is rational, it is automatically regular.   □

Remark. The two preceding propositions recover known results on the Albanese variety and on generalized Jacobians.

15. Maximal maps defined over k

We return to the notations and hypotheses of no. 13 and suppose that the variety V has the structure of k-*variety*. We can then consider maps $\alpha : V \to H$ defined over k (the group G and the homogeneous space H being themselves defined over k). Here, the distinction between group and homogeneous space becomes important, because H does not necessarily have a point rational over k.

A map $\alpha : V \to H$ defined over k will be called maximal if it is maximal over $\overline{k}$. Let L_k be the set of these maps. If α and α' are elements of L_k we will write $\alpha' \geq \alpha$ if there exists a morphism h, defined over k, such that $\alpha = h \circ \alpha'$.

The canonical map $L_k \to L$ is clearly increasing. More precisely:

Lemma 6. *The preorder relation of L_k is induced by that of L.*

PROOF. We must show that, if $\alpha : V \to H$ and $\alpha' : V \to H'$ are two elements of L_k such that there exists a morphism h (defined over $\overline{k}$) with $\alpha = h \circ \alpha'$, then h is defined over k. But, if σ denotes a k-automorphism of $\overline{k}$, then $V^\sigma = V$, $H^\sigma = H$, $H'^\sigma = H'$, $\alpha^\sigma = \alpha$ and $\alpha'^\sigma = \alpha'$. We deduce that $\alpha = h^\sigma \circ \alpha'$, and lemma 4 shows that $h^\sigma = h$, which means that h is defined over k. $\qquad\square$

In particular, the relation $\alpha \approx \alpha'$ in L_k is equivalent to the relation $\alpha \approx \alpha'$ in L. Thus there is no problem in considering L_k as a subset of L, and this is what we will do from now on.

Lemma 7. *In order that $\alpha \in L$ belong to L_k, it is necessary and sufficient that $\alpha^\sigma \approx \alpha$ for every k-automorphism σ of $\overline{k}$.*

PROOF. The condition is clearly necessary. We show that it is sufficient. There exists in any case a finite extension k_1 of k such that $\alpha \in L_{k_1}$ and we can suppose that k_1 is Galois over k; let $\mathfrak{g}$ be its Galois group. If $\sigma \in \mathfrak{g}$, we have by hypothesis $\alpha^\sigma \approx \alpha$. Denoting by H the k_1-homogeneous space associated to α, we thus have an isomorphism $h_\sigma : H \to H^\sigma$; according to lemma 6, this isomorphism is defined over k_1.

The formula $\alpha^\sigma = h_\sigma \circ \alpha$ gives $\alpha^{\sigma\tau} = (h_\sigma)^\tau \circ \alpha^\tau = (h_\sigma)^\tau \circ h_\tau \circ \alpha = h_{\sigma\tau} \circ \alpha$, so (lemma 4) $h_{\sigma\tau} = (h_\sigma)^\tau \circ h_\tau$. The theorem of *descent of the base field* (chap. V, no. 20, cor. 2 to prop. 12) then shows that H is k_1-isomorphic to a principal homogeneous space H_0 defined over k, which gives the desired result. $\qquad\square$

Lemma 8. *The set L_k is cofinal in L.*

PROOF. Let $\alpha \in L$, and choose as before a finite Galois extension k_1 of k such that $\alpha \in L_{k_1}$. Let $\mathfrak{g}$ be the Galois group of k_1/k, and let α^σ, $\sigma \in \mathfrak{g}$ be the conjugates of α. Putting $\beta = \mathrm{Sup}(\alpha^\sigma)$, we have $\beta \geq \alpha$ and as $\beta \approx \beta^\sigma$ for all $\sigma \in \mathfrak{g}$, lemma 7 shows that $\beta \in L_k$. $\qquad\square$

§4. Class field theory

In this §, k denotes a finite field with q elements, and V an irreducible k-variety. We write K for the field $k(V)$ of rational functions on V which are defined over k. We propose to determine the Galois group of the *maximal Abelian extension* of K.

16. Statement of the theorem

Let $\alpha : V \to H$ be an element of L_k, i.e. a maximal map defined over k. If G denotes the group associated to H, we write G_k (resp. H_k) for the set of points of G (resp. H) rational over k. According to cor. 1 to prop. 3, the set H_k is non-empty; it is thus a "principal homogeneous space" for G_k. Let $I_k(H)$ be the free Abelian group with base H_k; this group has a canonical surjective homomorphism $\epsilon : I_k(H) \to \mathbf{Z}$.

Let I_0 be the kernel of ϵ; an element $x \in I_0$ can be written as a formal linear combination $x = \sum n_i x_i$, $x_i \in H_k$, $n_i \in \mathbf{Z}$, $\sum n_i = 0$. To such an element x, we can associate the same sum, *computed in the group G_k*. Thus we get a surjective homomorphism $I_0 \to G_k$. If N is the kernel of this homomorphism, we put

$$H(k) = I_k(H)/N.$$

The homomorphism ϵ defines by passage to the quotient a surjective homomorphism $H(k) \to \mathbf{Z}$, which we denote again by ϵ, and the kernel of ϵ is identified with G_k. In other words, we have an exact sequence

$$0 \to G_k \to H(k) \xrightarrow{\epsilon} \mathbf{Z} \to 0. \tag{1}$$

The inverse image in $H(k)$ of the element $1 \in \mathbf{Z}$ is canonically identified with H_k (for its structure of principal homogeneous space for G_k).

[Of course, the preceding construction has nothing to do with the theory of algebraic groups. One can apply it to any principal homogeneous space for a commutative group: it is just the usual "barycentric calculus."]

Now let $\alpha' : V \to H'$ be an element of L_k such that $\alpha' \geq \alpha$ and let $h : H' \to H$ be the associated morphism. The map h defines a homomorphism $H'(k) \to H(k)$ and there is a commutative diagram

$$
\begin{array}{ccccccccc}
0 & \longrightarrow & G'_k & \longrightarrow & H'(k) & \longrightarrow & \mathbf{Z} & \longrightarrow & 0 \\
& & \ \downarrow{h_0} & & \ \downarrow{h} & & \ \downarrow{id.} & & \\
0 & \longrightarrow & G_k & \longrightarrow & H(k) & \longrightarrow & \mathbf{Z} & \longrightarrow & 0.
\end{array}
$$

According to lemma 4, the kernel of $h_0 : G' \to G$ is connected; cor. 2 to prop. 3 thus shows that $G'_k \to G_k$ is surjective, thus so is $H'(k) \to H(k)$.

Definition 3.   We call the projective limit of the groups $H_\alpha(k)$, where α runs through the preordered filtered set L_k, the cycle class group of V, and we write $A_k(V)$ for it.

Similarly, we will write $A_k^0(V)$ for the projective limit of the G_k. Because the G_k are *finite* groups, the projective limit of the exact sequences (1) is an exact sequence

$$0 \to A_k^0(V) \to A_k(V) \to \mathbf{Z} \to 0. \tag{2}$$

Note that the definition of a maximal map is *birational*: the preceding constructions and definitions do not depend on the *model V* chosen for the field K. For this reason, we will also write $A(K)$ and $A^0(K)$ in place of $A_k(V)$ and $A_k^0(V)$.

Now let Ω be an algebraic closure of K and let K_a be the *maximal Abelian extension* of K, i.e., the largest Abelian extension of K contained in Ω. It is well known that $\overline{k}$ is Abelian over k and its Galois group is $\hat{\mathbf{Z}}$, the completion of the group $\mathbf{Z}$ for the topology defined by subgroups of finite index. The map $\sigma : \lambda \to \lambda^q$ is a (topological) *generator* of this group. Since V was supposed *irreducible* (absolutely, i.e., over $\overline{k}$), the extensions $\overline{k}/k$ and K/k are linearly disjoint over k, which shows that the compositum $K\overline{k}$ is Abelian over K, with Galois group $\hat{\mathbf{Z}}$. In particular $K \subset K\overline{k} \subset K_a$. We denote by $\mathfrak{g}(K)$ the subgroup of the Galois group of K_a/K formed by elements which induce an element of $\mathbf{Z}$ (and not of $\hat{\mathbf{Z}}$) on $K\overline{k}$. We denote by $\mathfrak{g}^0(K)$ the Galois group of $K_a/K\overline{k}$. Thus there is an exact sequence

$$0 \to \mathfrak{g}^0(K) \to \mathfrak{g}(K) \to \mathbf{Z} \to 0. \tag{3}$$

We give $\mathfrak{g}^0(K)$ it natural topology as Galois group, which makes it a compact group. As for $\mathfrak{g}(K)$, it will be topologized by the condition that $\mathfrak{g}^0(K)$ be an open subgroup (in other words, its quotient $\mathbf{Z}$ should have the discrete topology). The group $\mathfrak{g}(K)$ is "almost" the Galois group of K_a/K (more precisely, the latter is the completion of $\mathfrak{g}(K)$ for the topology defined by the open subgroups of finite index).

We can now state the principal result of this chapter:

Theorem 1.   *There exists a canonical isomorphism from the exact sequence* (2) *to the exact sequence* (3).

(Of course, this isomorphism is the identity on $\mathbf{Z}$.)

One sees in particular from this that $\mathfrak{g}^0(K)$, which is the "geometric" Galois group, is isomorphic to the group $A^0(K)$ of cycle classes of degree 0.

The rest of this § is devoted to the proof of theorem 1. We begin by constructing, for every $\alpha \in L_k$, an extension $E_\alpha/K\overline{k}$ having the group $H_\alpha(k)$ as Galois group over K. We will see that, if $\alpha \geq \beta$, then $E_\alpha \supset E_\beta$, the corresponding homomorphism of Galois groups being the canonical map

$H_\alpha(k) \to H_\beta(k)$. Finally, we will show that every finite Abelian extension of K is contained in one of the fields E_α, which will clearly finish the proof.

17. Construction of the extensions E_α

Let $\alpha : V \to H$ be an element of L_k and let G be the corresponding group.

Proposition 14. *The pull-back by α of a separable isogeny $h : H' \to H$ is an irreducible Abelian covering of V.*

PROOF. Let $V' = \alpha^*(H')$; it is clear that V' is an Abelian covering of V, having as Galois group the kernel N of h. Let V'_1 be one of the irreducible components of V' and let N_1 be the subgroup of N formed by the elements $s \in N$ such that $s(V'_1) = V'_1$. Let $V'' = V'/N_1$ and $H'' = H'/N_1$; then $V'' = \alpha^*(H'')$ and, by construction, the covering $V'' \to V$ is trivial. Thus there exists a section $f : V \to V''$, or, what comes to the same thing, a map $\alpha'' : V \to H''$ lifting the map $\alpha : V \to H$. Since α is maximal, $H'' \to H$ is an isomorphism, whence $N_1 = N$, $V'_1 = V'$, and the covering V_1 is irreducible, as was to be shown. (Compare with the proof of prop. 10, no. 11.) $\square$

Now we choose a point $h \in H_k$ and define a map $\wp_k : G \to H$ by putting

$$\wp_k(x) = x^q - x + h.$$

Identifying H with G by taking h as the origin, this map is nothing other than the isogeny $\wp : G \to G$ of no. 6. Its pull-back W_h under α is a *covering of V, defined and Abelian over k*, and absolutely irreducible by the preceding proposition. Its Galois group is nothing other than the group G_k acting by translations. We will denote by K_h the field of rational functions on W_h and we put $E_h = K_h \overline{k}$. Because W_h is absolutely irreducible, K_h and $K\overline{k}$ are linear disjoint extensions of the field K, and it follows that E_h is a Galois extension of K, admitting for Galois group the product group $G_k \times \mathbf{Z}$ (here again, we mean the Galois group modified by replacing $\hat{\mathbf{Z}}$ with $\mathbf{Z}$).

Lemma 9. *The extension E_h/K does not depend on the choice of the point $h \in H_k$.*

PROOF. We must show that the coverings $W_h \to V$ and $W_{h'} \to V$ are *isomorphic over $\overline{k}$*, for any $h, h' \in H_k$. But, since the map $\wp : G \to G$ is surjective, there exists $c \in G$ such that $c^q - c = h - h'$; if $\theta : G \to G$ denotes translation by c, $\wp_h = \wp_{h'} \circ \theta$, which makes evident the desired isomorphism. $\square$

Note that, if $h \neq h'$, it is *impossible* to choose c rational over k, which shows that the extensions K_h and $K_{h'}$ are not isomorphic.

Definition 4. We denote by E_α the extension E_h/K, where h is any point of H_k.

(Lemma 9 shows that this definition is legitimate.)

It remains to establish a canonical isomorphism between the Galois group $\mathfrak{g}_\alpha$ of E_α/K and the group $H(k)$. Again we choose $h \in H_k$; identifying E_α with E_h, the group $\mathfrak{g}_\alpha$ is identified with the product $G_k \times \mathbf{Z}$, as we have seen. On the other hand, the choice of h also identifies $H(k)$ with $G_k \times \mathbf{Z}$, every element of $H(k)$ being written uniquely in the form $g + nh$, $g \in G_k$, $n \in \mathbf{Z}$. Whence we have an isomorphism $\rho_h : H(k) \to \mathfrak{g}_\alpha$.

Lemma 10. *The isomorphism ρ_h does not depend on the choice of the point $h \in H_k$.*

PROOF. It is clear that, if h and h' are two points of H_k, the isomorphisms ρ_h and $\rho_{h'}$ coincide on $G_k \subset H(k)$. Thus everything comes down to showing that ρ_h and $\rho_{h'}$ coincide on one element of degree 1 in $H(k)$, for example h. So put $\omega = \rho_h(h)$ and $\omega' = \rho_{h'}(h)$. We must show that ω and ω' coincide when $K_h\overline{k}$ and $K_{h'}\overline{k}$ are identified by means of the isomorphism $\theta : K_{h'}\overline{k} \to K_h\overline{k}$ introduced in the proof of lemma 9; in other words, we must establish the formula $\omega \circ \theta = \theta \circ \omega'$. But, by the definition of a pullback, W_h and $W_{h'}$ are subvarieties of $V \times G$, and the elements of $K_h\overline{k}$ and $K_{h'}\overline{k}$ can be interpreted as functions of two variables, $f(v, x)$, $v \in V$, $x \in G$, with values in $\overline{k}$. The operations θ, ω and ω' are explicitly given as

$$(\theta f)(v, x) = f(v, x + c).$$
$$(\omega f)(v, x) = f(v^{1/q}, x^{1/q})^q.$$
$$(\omega' f)(v, x) = f(v^{1/q}, x^{1/q} + h - h')^q.$$

Computing $(\omega\theta f)(v, x)$ and $(\theta\omega' f)(v, x)$, in both cases we find

$$f(v^{1/q}, x^{1/q} + c)^q,$$

which indeed shows that $\omega \circ \theta = \theta \circ \omega'$.   □

Lemma 10 implies:

Proposition 15. *The Galois group of E_α/K is canonically isomorphic to $H_\alpha(k)$.*

Remark. One can prove lemma 10 without computation, by using the reciprocity map (cf. §5).

Now let $\alpha' : V \to H'$ be another element of L_k with $\alpha' \geq \alpha$, and let $f : H' \to H$ be the morphism defined by α'.

Proposition 16. *If $\alpha' \geq \alpha$, the field $E_{\alpha'}$ contains the field E_α and there is a commutative diagram*

$$
\begin{array}{ccc}
H'(k) & \longrightarrow & \mathfrak{g}_{\alpha'} \\
\downarrow & & \downarrow \\
H(k) & \longrightarrow & \mathfrak{g}_\alpha.
\end{array}
$$

PROOF. Let $h' \in H'_k$ and $h = f(h')$. In view of the construction of E_α and of $E_{\alpha'}$ as well as the definition of the isomorphisms ρ_h and $\rho_{h'}$, it suffices to show that $K_h \subset K_{h'}$ and that the corresponding homomorphism of the Galois groups is the canonical homomorphism from G'_k to G_k. But, if $f_0 : G' \to G$ is the homomorphiosm associated to f, we have the commutative diagram

$$
\begin{array}{ccc}
G' & \xrightarrow{\ f_0\ } & G \\
\rho_{h'} \downarrow & & \downarrow \rho_h \\
H' & \xrightarrow{\ f\ } & H.
\end{array}
$$

This diagram shows that the pull-back of the isogeny $\wp_h$ by f is nothing other than the quotient of the isogeny $\wp_{h'}$ associated to the homomorphism $f_0 : G'_k \to G_k$ and our assertion follows immediately from that. □

18. End of the proof of theorem 1: first method

Let K' be the union of the extensions E_α, $\alpha \in L_k$. Propositions 15 and 16 show that K' is an Abelian extension of K containing $\overline{k}$ with (modified) Galois group $A(K)$, the cycle class group of V. We have $K' \subset K_a$ and everything comes down to showing that $K_a \subset K'$.

Let $\widehat{\mathfrak{g}}(K)$ be the Galois group of K_a/K; it is an extension of $\widehat{\mathbf{Z}}$ by $\mathfrak{g}^0(K)$. If σ denotes a generator of the group $\mathbf{Z}$, the choice of a representative of σ in $\widehat{\mathfrak{g}}(K)$ defines a section homomorphism $s : \mathbf{Z} \to \widehat{\mathfrak{g}}(K)$. This homomorphism extends by continuity to $\widehat{\mathbf{Z}}$ since $\widehat{\mathfrak{g}}(K)$ is a compact totally disconnected group. Thus $\widehat{\mathfrak{g}}(K)$ decomposes (non-canonically, of course) as $\mathfrak{g}^0(K) \times \widehat{\mathbf{Z}}$, which translates to a decomposition $K_a = L_0\overline{k}$, with L_0 linearly disjoint from $\overline{k}$. Since $\overline{k} \subset K'$, it will suffice to show that $L_0 \subset K'$.

Thus let L/K be a finite extension contained in L_0. Since this extension is linearly disjoint from $\overline{k}$, it corresponds to an Abelian covering $\pi : W \to V$, where W is an irreducible k-variety. Let N be its Galois group. According to the corollary to proposition 7 of §2, the covering W is the pull-back over the field k of a separable isogeny $G' \to G$ by a rational map $f : V \to G$. According to lemma 3, we can factor f as

$$
V \xrightarrow{\ \alpha\ } H_\alpha \xrightarrow{\ \varphi\ } G
$$

where φ is a morphism, and α is a maximal map; according to lemma 8, we can choose α to be in L_k. If σ is a k-automorphism of $\overline{k}$, then $f = f^\sigma$, so $\varphi \circ \alpha = \varphi^\sigma \circ \alpha$, which shows that $\varphi = \varphi^\sigma$, in other words that φ is defined over k. Let $H' \to H_\alpha$ be the pull-back by φ of the isogeny $G' \to G$. Since φ is defined over k, the same is true of this isogeny and its kernel N is formed by elements rational over k (since this is true for $G' \to G$). We have $\alpha^*(H') = W$ and as W is irreducible, so is H'. Let h' be a point of H' rational over k; such a point exists by corollary 1 to proposition 3; let h be its image in H_α. Taking these points for the origin, we can identify H' and H_α with groups and apply proposition 6 of §1 to them. We deduce that $H' \to H_\alpha$ is a quotient covering of the covering $\wp_h : G_\alpha \to H_\alpha$ (G_α denoting the group associated to H_α). The covering $W = \alpha^*(H')$ is thus a quotient of the covering W_h, itself a pull-back of the isogeny $\wp_h$, which shows that $L \subset K_h$, so $L \subset E_\alpha$, which finishes the proof. $\square$

Remark. Because the group $G(N)$ of prop. 7 is a *linear* group, in the preceding proof we could have used only linear groups. Thus, *the projective limit of the $H_\alpha(k)$ does not change when one restricts it to maximal maps $\alpha \in L_k$ corresponding to linear groups*. This is all the more curious because these maximal maps *do not form* a cofinal system in L_k.

19. End of the proof of theorem 1: second method

First we make the following definition:

Definition 5. Let $\alpha : V \to H_\alpha$ be an element of L and let F be a finite extension of $K\overline{k}$. One says that F is of type α if F is the pull-back of a separable isogeny $H' \to H_\alpha$.

This is a "geometric" notion (i.e., it is relative to the structure of "variety" and not of "k-variety"). If we take for α the canonical map of V to its Albanese variety, we get the notion of an extension of "Albanese type" introduced by Lang [49].

Theorem 1 is a consequence of the two more precise propositions below:

Proposition 17. *For every finite Abelian extension F of $K\overline{k}$, there exists $\alpha \in L$ such that F is of type α.*

PROOF. Here again, this is a "geometric" statement. One proves it by reasoning analogous to that of the preceding no. (but simpler): applying the corollary to prop. 7 of §2, we see that F is the pull-back of a separable isogeny by a rational map $f : V \to G$ (instead of applying this corollary, one can invoke the theories of Kummer and Artin-Schreier, since $\overline{k}$ contains all the roots of unity). One has then only to factor f as $V \xrightarrow{\alpha} H_\alpha \to G$ where α is maximal (cf. lemma 3). $\square$

Now if E/K is a finite extension, we say that E/K *is of type* α if this is true of $E\overline{k}/K\overline{k}$. With this convention, we have:

Proposition 18. *Let* $\alpha \in L_k$. *In order that an Abelian extension* E/K *be of type* α, *it is necessary and sufficient that it be contained in* E_α.

(Remember that E is assumed *Abelian over* K!)

PROOF. The condition is clearly sufficient. We show that it is necessary. The compositum $E\overline{k}$ is an Abelian extension of K (as the compositum of two Abelian extensions). The argument of the preceding no. for K_a thus shows that $E\overline{k} = E'\overline{k}$ where E' is an extension of K linearly disjoint from $\overline{k}$, which thus corresponds to an irreducible covering $\pi : W \to V$. It will suffice to show that E' is contained in E_α.

By hypothesis, the covering W is of the form $\alpha^*(H')$, where $H' \to H_\alpha$ is an isogeny of H_α (this being true a priori over $\overline{k}$ and not over k). Let k_1 be a finite extension of k such that this is true over k_1, and let $\mathfrak{g}$ be the Galois group of k_1/k. If σ denotes an element of $\mathfrak{g}$, the covering W is the pull-back of the covering H'^σ. But because α is maximal, two isogenies of H_α which have the same pull-back by α^* are isomorphic (cf. the argument of prop. 11). Thus we get an isomorphism $\varphi_\sigma : H' \to H'^\sigma$, clearly unique, thus defined over k_1. One immediately checks the relation $(\varphi_\sigma)^\tau \circ \varphi_\tau = \varphi_{\sigma\tau}$, which permits us to descend the base field of H' from the field k_1 to the field k. Then applying, as in the preceding no., prop. 6 of §1, we deduce that E' is contained in an extension K_h, whence the desired result. $\square$

[We indicate a variant of this proof, which does not use descent of the base field:

Let k_1 be, as above, a finite extension of k such that W is of the form $\alpha^*(H')$, with $H' \to H_\alpha$ an isogeny over k_1. We conclude in any case that E/k_1 is contained in the extension E'_α/Kk_1 corresponding to the maximal map α and to the base field k_1. Thus everything comes down to showing that *the largest Abelian extension of K contained in E'_α is E_α*. This is a question of Galois groups: we have the "tower" of fields $k \subset k_1 \subset E_\alpha \subset E'_\alpha$, and we must show that the Galois group $\mathfrak{t}$ of E'_α/E_α is contained in the commutator subgroup of the Galois group G of E'_α/K. This last group is itself an extension of $\mathfrak{g}$ by $H_\alpha(k')$. A trivial direct computation shows that the interior automorphisms defined by the elements of $\mathfrak{g}$ operate on $H_\alpha(k')$ in the obvious way. It follows that the commutator subgroup of G contains the subgroup of $H_\alpha(k')$ generated by the $x - x^q$, $x \in H_\alpha(k')$. On the other hand, $\mathfrak{t}$ is equal to the kernel of the canonical homomorphism $H_\alpha(k') \to H_\alpha(k)$, which is given explicitly by the formula

$$x \to x + x^q + \cdots + x^{q^{d-1}} = \text{Tr}(x), \qquad \text{with } d = [k_1 : k].$$

To show that $\mathfrak{t}$ is contained in the commutator subgroup of G, it suffices to prove that every element $x \in H_\alpha(k')$ with $\text{Tr}(x) = 0$ can be written in the form $y - y^q$, which follows from the fact that the map $y \to y - y^q$ is surjective.]

20. Absolute class fields

The interest of proposition 18 is that it permits one to determine the type of an Abelian extension by *geometric* means (i.e., working over $\overline{k}$). We treat, by way of example, the case of unramified extensions:

Let V be a *non-singular projective* variety and let $\alpha : V \to A$ be the canonical map of V into its Albanese variety. Since V is non-singular, α is everywhere regular and it is clear that every covering of V of type α (or, as one says, "of Albanese type") is *Abelian and unramified*. We ask if the converse is true. One reduces immediately to the following two cases:

a) *Cyclic covering of order n prime to p.*

Kummer theory shows that these coverings correspond to an element d of order n of the group $C(V)$ of divisor classes of V for linear equivalence, cf. [77], no. 15. The group $C(V)$ contains the subgroup $P(V)$ of divisor classes algebraically equivalent to zero; the quotient

$$N(V) = C(V)/P(V)$$

is the *Neron-Severi group*. The group $P(V)$ is isomorphic to the underlying group of the *Picard variety* of V, or, what is the same thing, of the dual variety of A (cf. Lang [52], chap. VI). We have $P(A) = P(V)$, and the group $N(A)$ has no torsion (cf. Barsotti [5], or [78]). It follows that the element d is of type α if and only it belongs to $P(V)$. For this to be so for all elements of $C(V)$ of order prime to p, it is necessary and sufficient ($P(V)$ being divisible) that $N(V)$ contain no non-zero element of order prime to p.

b) *Cyclic covering of order a power of p.*

We examine the case where the order is p. Artin-Schreier theory (cf. [77], no. 16) shows that such a covering corresponds to an element $x \in H^1(V, \mathcal{O}_V)$ left fixed by the Frobenius operation F. This covering is of Albanese type if and only if x is in the image of the homomorphism $\alpha^* : H^1(A, \mathcal{O}_A) \to H^1(V, \mathcal{O}_V)$; this homomorphism is moreover injective, cf. chap. VII, as well as [78], no. 9. Denoting by $H^1(A, \mathcal{O}_A)_s$ the *semi-simple part* of $H^1(A, \mathcal{O}_A)$ ([77], *loc. cit.*), we see that α^* must map $H^1(A, \mathcal{O}_A)_s$ onto $H^1(V, \mathcal{O}_V)_s$. Conversely, if this condition is satisfied, one shows that every cyclic covering of order p^n of V which is unramified is of Albanese type (arguing by induction on n, using the fact that every covering of Albanese type which is cyclic of order p^{n-1} is the image of a covering of the same type cyclic of order p^n). Finally, we have:

Proposition 19. *In order that every unramified Abelian extension of V be of Albanese type, it is necessary and sufficient that the Néron-Severi* *group of V have no torsion of order prime to p, and that the map α^* : $H^1(A, \mathcal{O}_A)_s \rightarrow H^1(V, \mathcal{O}_V)_s$ be surjective.*

In this case, proposition 18 shows that *the maximal unramified extension of K has (modified) Galois group $H_\alpha(k)$*, i.e., an extension of $\mathbf{Z}$ by the group A_k of points of A rational over k. This is the analog of the *absolute class field* of number theory.

Remarks. 1) The condition of prop. 19 holds *if the Néron- Severi group of V has no torsion and if $h^{0,1}(V) = \dim A$* (we put, as usual,

$$ h^{0,1}(V) = \dim \ H^1(V, \mathcal{O}_V)). $$

Indeed, $\dim H^1(A, \mathcal{O}_A) = \dim A$ (cf. chap. VII), and the map α^* is thus surjective. In particular, these two conditions are satisfied if V is a curve.

2) In the general case, one can show that a cyclic extension of order n is of Albanese type if and only if, for every $m \geq 1$, it is contained in an unramified cyclic extension of degree n^m (defined over $\overline{k}$). When n is prime to p, that can be seen by using the fact that the Néron-Severi group is of finite type. When n is a power of p, this can be deduced from **[77]**, *loc. cit.*, combined with a result of Mumford **[117]**, p. 196.

21. Complement: the trace map

Let E be a finite Galois extension of K and let $\mathfrak{g}$ be its Galois group. Denoting by E_α the maximal Abelian extension of E, we have $K_a E \subset E_a$ and, on the other hand, $K_a \bigcap E$ is the largest Abelian extension of K contained in E (cf. the diagram below).

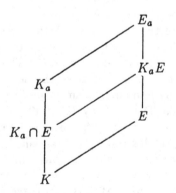

Passing to Galois groups and using thm. 1, we deduce the exact sequence

$$ A(E) \rightarrow A(K) \rightarrow \mathfrak{g}/\mathfrak{g}' \rightarrow 0, $$

where $\mathfrak{g}/\mathfrak{g}'$ denotes the quotient of $\mathfrak{g}$ by its commutator subgroup.

We will return in §7 to this exact sequence. The homomorphism $A(E) \to A(K)$ which figures there corresponds, in the classical case, to the *trace* operation; we designate it also by Tr.

One can try to determine Tr more or less explicitly. There are two cases to consider, that where $E = Kk'$, k'/k being a finite extension, and that where E/K is linearly disjoint from $\overline{k}$ (a "geometric" extension, as Lang says).

First case: $E = Kk'$. One can define $A(E)$ as the projective limit, for $\alpha \in L_k$, of the groups $H_\alpha(k')$, since one knows that L_k is cofinal in L. But, for each $\alpha \in L_k$, there is a trace homomorphism:

$$\text{Tr} : H_\alpha(k') \to H_\alpha(k)$$

defined by the formula given above:

$$\text{Tr}(x) = x + x^q + \cdots + x^{q^{d-1}}, \qquad \text{with } d = [k' : k].$$

By passage to the limit, these homomorphisms define the sought after homomorphism $Tr : A(E) \to A(K)$.

Note that this homomorphism *multiplies the degrees by d*.

Second case: E is a geometric extension. Let $\{x_{\alpha'}\}$ be an element of $A(E)$, for α' running through the set L'_k of maximal maps of a model V' of E. To associate to $x = \{x_{\alpha'}\}$ an element $\text{Tr}(x) \in A(K)$, we must define $Tr(x)_\alpha \in H_\alpha(k)$ for all $\alpha \in L_k$. But, composing $W \to V \xrightarrow{\alpha} H_\alpha$, we find a map which factors as $W \xrightarrow{\alpha'} H' \xrightarrow{\varphi} H_\alpha$ where α' and φ can be defined over k. Putting

$$Tr(x)_\alpha = \varphi(x_{\alpha'})$$

one checks that this is the desired homomorphism.

Note that this homomorphism *conserves the degrees*.

§5. The reciprocity map

The hypotheses and notations are the same as those of §4.

22. The Frobenius substitution

Let P be a point of V algebraic over k and let $d = [k(P) : k]$. The point P has d conjugates over k, which are:

$$P, P^q, \ldots, P^{q^{d-1}}.$$

The sum $\mathfrak{p}$ of the conjugates of P is a *prime cycle rational over k*; the integer d is its *degree*. The local rings of the points P^{q^i} coincide; we will denote them either by $\mathcal{O}_P$ or by $\mathcal{O}_{\mathfrak{p}}$. Similarly we will write $k(\mathfrak{p})$ in place of $k(P)$.

Now let L/K be a finite Galois extension of degree n and let $\mathfrak{g}_{L/K}$ (or simply $\mathfrak{g}$) be its Galois group. Suppose that P *is not ramified in* L. If we denote by $\mathcal{O}'_{\mathfrak{p}}$ the integral closure of $\mathcal{O}_{\mathfrak{p}}$ in L, the semi-local ring $\mathcal{O}'_{\mathfrak{p}}$ decomposes as the intersection of local rings $\mathcal{O}'_{\mathfrak{p}_i}$ $(i = 1, \ldots, r)$. Let $\mathfrak{m}_{\mathfrak{p}}$ be the maximal ideal of $\mathcal{O}_{\mathfrak{p}}$; one knows that $\mathcal{O}'_{\mathfrak{p}}/\mathfrak{m}_{\mathfrak{p}}\mathcal{O}'_{\mathfrak{p}}$ is a semi-simple algebra of dimension n over $k(\mathfrak{p})$, with Galois group $\mathfrak{g}$. This algebra decomposes into a product of fields $k(\mathfrak{p}_i)$ which are just the residue fields of the $\mathcal{O}'_{\mathfrak{p}_i}$. Let $\mathfrak{g}_i$ be the subgroup of $\mathfrak{g}$ formed by the elements that leave $\mathcal{O}'_{\mathfrak{p}_i}$ stable. This group is identified with the Galois group of the extension $k(\mathfrak{p}_i)/k(\mathfrak{p})$. Thus it is a cyclic group of order $f = [k(\mathfrak{p}_i) : k(\mathfrak{p})]$ generated by an element $(\mathfrak{p}_i, L/K)$ which corresponds to raising to the power q^d in $k(\mathfrak{p}_i)/k(\mathfrak{p})$. The element $(\mathfrak{p}_i, L/K)$ is called *the Frobenius substitution at* $\mathfrak{p}_i$ *in* L/K. It is an element of the Galois group $\mathfrak{g}_{L/K}$ whose order is equal to $f = n/r$.

Changing $\mathfrak{p}_i$ to $\mathfrak{p}_j$ changes $(\mathfrak{p}_i, L/K)$ to a conjugate element. When the group $\mathfrak{g}_{L/K}$ is *Abelian* (which is the most important case in what follows), we thus see that $(\mathfrak{p}_i, L/K)$ does not depend on i, and we can denote it by $(\mathfrak{p}, L/K)$; one then calls it *the Frobenius substitution at* $\mathfrak{p}$ *in* L/K. We will also write $(P, L/K)$ in place of $(\mathfrak{p}, L/K)$.

In order that $(\mathfrak{p}, L/K) = 0$ it is necesssary and sufficient that $\mathfrak{p}$ be *completely decomposed* in L/K, i.e., $f = 1$ or $k(\mathfrak{p}_i) = k(\mathfrak{p})$ for all i. The "functorial" properties of the Frobenius substitution are the same as in number theory (cf. [30], II, §1). We make explicit only the following:

(*Transitivity*). If $K \subset E \subset L$, L being a finite Abelian extension of K, the image of $(\mathfrak{p}, L/K)$ in $\mathfrak{g}_{E/K}$ is equal to $(\mathfrak{p}, E/K)$.

(We assume, of course, that $\mathfrak{p}$ is unramified in L/K.)

This property permits one to define the symbol $(\mathfrak{p}, L/K)$ when L is an *infinite* Abelian extension of K; we will see an example of this later.

23. Geometric interpretation of the Frobenius substitution

We are going to succesively examine the case where L comes from an extension of the base field and the case where L is "geometric".

i)$L = Kk'$, *where* k'/k *is an extension of degree* n.

In this case the algebra $\mathcal{O}'_{\mathfrak{p}}/\mathfrak{m}_{\mathfrak{p}}\mathcal{O}'_{\mathfrak{p}}$ is identified with $k(\mathfrak{p}) \otimes_k k'$. We deduce that $(\mathfrak{p}, L/K)$ is the element of the Galois group $\mathfrak{g} = \mathfrak{g}_{k'/k} = \mathbf{Z}/n\mathbf{Z}$ which is congruent to $d \bmod n$.

Passing to the limit over k', we get the more suggestive formula

$$(\mathfrak{p}, K\overline{k}/K) = \deg(\mathfrak{p}).$$

ii)L *is linearly disjoint from* $K\overline{k}$.

In this case, L corresponds to a covering $W \to V$, irreducible over $\bar{k}$. The inverse image in W of the cycle $\mathfrak{p}$ is a cycle rational over k, which decomposes into a sum of rational prime cycles

$$\mathfrak{p} = \sum \mathfrak{p}_i \quad \text{(in } W\text{)}.$$

Choosing a point $P \in \mathfrak{p}$, the inverse image of P in W decomposes into classes of conjugate points, and each class gives rise to exactly one of the $\mathfrak{p}_i$. If $Q \in \mathfrak{p}_i$ projects to P, the same is true of the point Q^{q^d}, since $P^{q^d} = P$. Thus there exists a unique $\sigma \in \mathfrak{g}$ such that $\sigma(Q) = Q^{q^d}$. This element σ *is the Frobenius substitution* $(Q, L/K) = (\mathfrak{p}_i, L/K)$. Indeed, if $f \in \mathcal{O}_Q$, we have $f^\sigma(Q) = f(\sigma.Q) = f(Q^{q^d}) = f(Q)^{q^d}$, which shows that σ induces, by passage to the quotient, raising to the q^d-th power.

Note that "$\mathfrak{p}$ decomposes completely in L" is equivalent to "$k(Q) = k(P)$ for all Q projecting to P."

24. Determination of the Frobenius substitution in an extension of type α

Let $\alpha : V \to H$ be an element of L_k and let $\mathfrak{p} = P + \cdots + P^{q^{d-1}}$ be a rational prime cycle on V, *such that α is regular at P*. This last condition implies that $\mathfrak{p}$ is unramified in the extension E_α/K associated to α (cf. no. 17). One can thus speak of $(\mathfrak{p}, E_\alpha/K)$ which is an element of the Galois group $\mathfrak{g}_{E_\alpha/K}$. But this Galois group has been determined (*loc. cit.*): it is the completion (for the topology of subgroups of finite index) of the group $H_\alpha(k)$. We are going to see that, in fact, $(\mathfrak{p}, E_\alpha/K)$ belongs to $H_\alpha(k)$. More precisely, we put

$$\alpha(\mathfrak{p}) = \sum_{i=0}^{i=d-1} \alpha(P^{q^i}),$$

the sum being computed in the group $H_\alpha(\bar{k})$ defined in no. 13. Because $\mathfrak{p}$ is invariant by every k-automorphism of $\bar{k}$, the same is true of $\alpha(\mathfrak{p})$, which shows that $\alpha(\mathfrak{p}) \in H_\alpha(k)$. This being the case, the result that we have in mind can be stated as follows:

Theorem 2. $(\mathfrak{p}, E_\alpha/K) = \alpha(\mathfrak{p})$, *where we identify $H_\alpha(k)$ with an everywhere dense subgroup of $\mathfrak{g}_{E_\alpha/K}$ as stated above.*

PROOF. In light of the transitivity of the Frobenius substitution, it suffices to show that $\alpha(\mathfrak{p})$ induces $(\mathfrak{p}, E_\alpha/K)$ for the subextensions E of E_α generating E_α.

First taking $E = K\bar{k}$, we know that $(\mathfrak{p}, E/k) = \deg(\mathfrak{p})$, and on the other hand, the image of $\alpha(\mathfrak{p})$ in $\mathbf{Z}$ is evidently also $\deg(\mathfrak{p})$, whence the result.

Next taking $E = K_h$, where $h \in H_k$, the group $\mathfrak{g}_{E/K}$ is G_k (denoting by G the group associated to H), the homomorphism $H_\alpha(k) \to G_k$ being that which associates to $\sum n_i h_i$ the element $\sum n_i(h_i - h)$ of G_k. In particular, the image of $\alpha(\mathfrak{p})$ is

$$\sigma = \sum_{i=0}^{i=d-1} \alpha(P)^{q^i} - dh.$$

Let us show that $\sigma = (\mathfrak{p}, E/K)$, i.e., that σ transforms a point Q projecting to P to the point Q^{q^d}. The covering W_h corresponding to K_h is a subvariety of $V \times G$. The point Q can thus be identified with a pair (P, Q_1), where $Q_1 \in G$ satisfies

$$Q_1^q - Q_1 + h = \alpha(P).$$

Raising this identity to the q^i-th power $(i = 0, \ldots, d-1)$, we get

$$Q_1^{q^{i+1}} - Q_1^{q^i} + h = \alpha(P)^{q^i}.$$

Summing the d identities thus obtained, we find

$$Q_1^{q^d} - Q_1 + dh = \sum_{i=0}^{i=d-1} \alpha(P)^{q^i},$$

whence

$$Q_1^{q^d} - Q_1 = \sigma.$$

Thus $\sigma(Q) = (P, Q_1 + \sigma) = (P, Q_1^{q^d}) = Q^{q^d}$, which shows that $\sigma = (\mathfrak{p}, E/K)$. As the extensions $K\bar{k}/K$ and K_h/K generate E_α, theorem 2 is proved.   □

Corollary. *Let $\wp : G \to G$ be the isogeny $x \to x^q - x$, and let $\mathfrak{p}$ be a rational prime cycle of G. The Frobenius substitution of $\mathfrak{p}$ in the covering $\wp$ is equal to the element of G_k which is the sum (in G) of the points of $\mathfrak{p}$.*

In particular, if $d = 1$, i.e., if $\mathfrak{p}$ is a single point $P \in G_k$, we see that *the Frobenius substitution at P* (considered as prime cycle) *is P itself* (considered as an element of the Galois group).

Remark. The argument of theorem 2 applies more generally to any covering $W \to V$ which is the pull-back of an isogeny $G' \to H$. The fact that $\alpha : V \to H$ was assumed maximal was not used in any essential way.

25. The reciprocity map: statement of results

Let, as before, L/K be a finite Abelian extension and let V be a model of the field K for which no point ramifies in L (such a model always exists: it suffices to take any model, and to remove the points of ramification, since one knows that these form a proper algebraic subset). If $Z_k(V)$ denotes the group of cycles of V rational over k, this group admits as basis the set of rational prime cycles; the map $\mathfrak{p} \to (\mathfrak{p}, L/K)$ then defines by linearity a homomorphism from $Z_k(V)$ to $\mathfrak{g}_{L/K}$ called the *reciprocity map*.

Theorem 3. *For every model V of K containing no points ramified in L, the reciprocity map $Z_k(V) \to \mathfrak{g}_{L/K}$ is surjective.*

We will prove this result a little later. First we are going to give several equivalent formulations:

First let $\alpha : V \to H$ be an element of L_k; we can extend the map $\mathfrak{p} \to \alpha(\mathfrak{p}) \in H_\alpha(k)$ to a homomorphism $c \to \alpha(c)$ from $Z_k(V)$ to $H_\alpha(k)$.

Theorem 3′. *For every model V of K such that α is defined at every point of V, the homomorphism $Z_k(V) \to H_\alpha(k)$ is surjective.*

Let us admit theorem 3 and let H' be the image of $Z_k(V)$ in $H_\alpha(k)$. Suppose that $H' \neq H_\alpha(k)$. As $H_\alpha(k)$ is an Abelian group of finite type, there exists a subgroup H'' of $H_\alpha(k)$ of finite index $n > 1$ in $H_\alpha(k)$ and containing H'. This subgroup corresponds to a finite extension L/K of degree n, in which the reciprocity map is trivial (theorem 2), which contradicts theorem 3. Conversely, if we admit theorem 3′ for all $\alpha \in L_k$ (or only for a cofinal family of such α), every finite extension L/K is contained in E_α/K, whence immediately theorem 3.

[Theorem 3′ justifies the terminology "cycle class group" for the projective limit $A(K)$ of the $H_\alpha(k)$; indeed it shows that $H_\alpha(k)$ is the quotient of the group $Z_k(V)$ of rational cycles on V by the equivalence relation defined by α. If, for example, $\alpha : V \to J$ is the canonical map of a curve into its Jacobian, this group is just the group of divisor classes of V in the sense of linear equivalence, cf. §6.]

Theorem 3″. *Let F/K be an arbitrary finite extension (not necessarily Galois) and let V be a model of K such that every rational prime cycle of V is unramified and completely decomposed in F. Then $F = K$.*

Let r be an integer ≥ 1. Denote by 3_r (resp. $3_r''$) the statement of theorem 3 (resp. theorem 3″) for all fields K of transcendence degree r over k; we similarly denote by $3_r''{-}G$ and $3_r''{-}A$ the variants of theorem 3″ where the extension F/K is assumed to be Galois or Abelian. We show equivalences:

$$3_r'' \Longleftrightarrow 3_r''{-}G \Longleftrightarrow 3_r''{-}A \Longleftrightarrow 3_r .$$

Let F/K be an extension satisfying the hypotheses of theorem 3″, and let F'/K be the smallest Galois extension containing it. Because F' is the compositum of F and of its conjugates, F'/K satisfies the same hypotheses. If $[F : K] > 1$, then $[F' : K] > 1$ and the Galois group $\mathfrak{g}_{F'/K}$ contains a non-trivial cyclic subgroup $\mathfrak{g}''$. Let K'' be the subfield corresponding to $\mathfrak{g}''$.

The extension F'/K'' also satisfies the hypotheses of theorem 3″, whence $3_r''$—$A \implies 3_r''$. As the other implications $3_r'' \implies 3_r''$—$G \implies 3_r''$—A are trivial, these three assertions are equivalent.

Now let L/K be an Abelian extension with Galois group $\mathfrak{g}$, and let $\mathfrak{g}' \subset \mathfrak{g}$ be the image of the reciprocity map. If K' is the subfield corresponding to $\mathfrak{g}'$, the transitivity of the Frobenius substitution shows that $(\mathfrak{p}, K'/K) = 0$ for all $\mathfrak{p}$, and $\mathfrak{p}$ is completely decomposed in K'. Admitting $3_r''$—A, one thus has $K' = K$, whence $\mathfrak{g}' = \mathfrak{g}$, which proves 3_r. The implication $3_r \implies 3_r''$—A is immediate.

26. Proof of theorems 3, 3′, and 3″ starting from the case of curves

We will give in §6 a direct proof of theorem 3′ in the particular case where V is a *curve* ($r = 1$). We are going to show how one can pass from this to the general case, arguing by induction on r. We will use form $3_r''$—A of the theorem. In other words, we will give ourselves an Abelian extension F/K, with Galois group $\mathfrak{g}$ satisfying the hypotheses of thm. 3″, and we will show that necessarily $F = K$. If k' denotes the algebraic closure of k in F, we can reduce to studying the two extensions F/Kk' and Kk'/K. The problem thus divides in two:

i) *Case where $F = Kk'$, with $[k' : k] = n$.*

We must show that, if $n > 1$, the variety V *contains a rational cycle of degree $\not\equiv 0 \mod n$*. After replacing V by one of its open sets, we can assume that V is a locally closed subvariety of a projective space $\mathbf{P}_m$. We are going to apply "Bertini's Theorem" to the injection $V \to \mathbf{P}_m$.

Lemma 11. *Let $f : V \to \mathbf{P}_m$ be a regular map of an irreducible variety V into a projective space and suppose that $\dim f(V) \geq 2$. In the dual projective space $\mathbf{P}_m{}^*$, there exists an algebraic subset Y, distinct from $\mathbf{P}_m{}^*$, such that for every hyperplane $E \in \mathbf{P}_m{}^* - Y$, the set $f^{-1}(E)$ is an irreducible subvariety of V of dimension $r - 1$.*

PROOF. We recall the principle of the proof: thanks to a lemma of field theory (cf. [51], p. 213) one shows that $f^{-1}(E)$ is irreducibe provided that E is a generic point of $\mathbf{P}_m{}^*$. One then proves, using Chow coordinates, that the condition "$f^{-1}(E)$ is irreducible" is algebraic in E. For more details, see Zariski [101] or Matsusaka [56]. □

Thus let Y be the subspace of $\mathbf{P}_m{}^*$ whose existence is asserted by the preceding lemma; Y is defined over $\overline{k}$. Choose an integer $s > 1$ prime to $n = [k' : k]$ and let k_s be the extension of k composed of all the extensions of degree a power of s. We have $k \subset k_s \subset \overline{k}$ and the field k_s has infinitely many elements. According to an elementary result (cf. Bourbaki, Alg. IV,

§2, no. 5), there exists a hyperplane $E \in \mathbf{P}_m{}^*$ which is rational over k_s and does not belong to Y. Putting $V' = V \cap E$, the variety V' is irreducible, and defined over an extension k'' of k of degree a power of s. According to the induction hypothesis, V' contains a cycle rational over k'' of degree $d \not\equiv 0 \bmod n$. If c' denotes this cycle, the sum of c' and its conjugates over k is a cycle c of V, rational over k, and of degree $s^a d \not\equiv 0 \bmod n$, whence the desired result in this case.

ii) *F and $K\overline{k}$ are linearly disjoint over K.*

The extension F/K then corresponds to a covering $W \to V$; we can suppose as before that V is embedded in a projective space $\mathbf{P}_m$. Applying lemma 11 to $W \to \mathbf{P}_m$, we see that there exists a hyperplane E defined over a finite extension k' of k such that the inverse image W' of $V' = V \cap E$ is an (absolutely) irreducible k'-variety. The degree of the covering $W' \to V'$ is the same as that of the covering $W \to V$ (note that one can always assume that $W \to V$ is unramified, after restricting V). If this degree is > 1, the induction hypothesis shows that there exists a pair (P', Q'), $P' \in V'$, $Q' \in W', Q'$ projecting to P', such that $k'(P') \neq k'(Q')$. A fortiori, $k(P') \neq k(Q')$, which shows that P' does not decompose completely in L/K and finishes the proof. $\qquad\square$

27. Kernel of the reciprocity map

Let $\alpha : V \to H_\alpha$ be a maximal map defined over k and suppose as before that α is everywhere regular on V. Let L/K be a finite Abelian extension of type α, i.e., contained in E_α/K. Its Galois group $\mathfrak{g}$ is a quotient $H_\alpha(k)/N$, where N denotes a subgroup of finite index in $H_\alpha(k)$. Let k' be the algebraic closure of k in L; the normalization W of V in L/K is a k'-variety, whose cycle group $Z_{k'}(W)$ is well defined. By composition,

$$Z_{k'}(W) \to Z_{k'}(V) \xrightarrow{\mathrm{Tr}} Z_k(V),$$

we get a homomorphism from the cycle group of W ito that of V. This homomorphism plays the role of a trace and we will denote it by Tr.

Proposition 20. *The kernel of the reciprocity homomorphism $Z_k(V) \to \mathfrak{g}$ is generated by the kernel of $\alpha : Z_k(V) \to H_\alpha(k)$ and by the image of $Tr : Z_{k'}(W) \to Z_k(V)$.*

PROOF. It comes to the same thing to say that the image of $Z_{k'}(W)$ by the composed map $Z_{k'}(W) \to Z_k(V) \to H_\alpha(k)$ is equal to N. In any case it is clear that N contains this image (a trace always belongs to the kernel of the reciprocity map), and to see that equality holds we are reduced to the following two particular cases:

i) $L = Kk'$, with k' a finite extension of k.

We must show that in this case $Z_{k'}(V)$ has for image in $H_\alpha(k)$ the set I_d of elements of degree divisible by $d = [k' : k]$. But we can factor the homomorphism $Z_{k'}(V) \to H_\alpha(k)$ as

$$Z_{k'}(V) \to H_\alpha(k') \xrightarrow{\text{Tr}} H_\alpha(k).$$

According to thm. 3', the first homomorphism is surjective and, according to the corollary to prop. 5 of §1, the second has I_d for image, whence the result.

ii) *L is linearly disjoint from $K\overline{k}$.*

The extension L/K then corresponds to a covering $W \to V$, which is the inverse image of an isogeny $H' \to H_\alpha$, which is itself a quotient of an isogeny $\wp_h : G \to H_\alpha$. The group N can then be characterized as the image of $H'(k)$ in $H_\alpha(k)$ and everything comes down to seeing that $\alpha' : Z_k(W) \to H'(k)$ is surjective. This is a consequence of theorem 3' and the following lemma:

Lemma 12. *The map $\alpha' : W \to H'$ is maximal.*

PROOF. Let $H'' \to H'$ be an isogeny such that α' lifts to $s : W \to H''$; we show that this isogeny is trivial. If $\mathfrak{g}$ denotes the kernel of $H' \to H_\alpha$, which is also the Galois group of the covering $W \to V$, the image of $s(\sigma.w)$ in H' ($\sigma \in \mathfrak{g}, w \in W$) is equal to $\alpha'(w) + \sigma$. We conclude that $w \to s(\sigma.w) - s(w)$ takes its values in the kernel of $H'' \to H$, thus is a constant map equal to σ''. The map $\sigma \to \sigma''$ is a homomorphism from $\mathfrak{g}$ to the group G'' corresponding to H''. Putting $H''' = H''/\mathfrak{g}$, the map s defines by passage to the quotient a map $t : V \to H'''$. In light of the maximal character of α, this implies that $H'' \to H$ is an isomorphism and the same is true of $H'' \to H'$, which finishes the proof.   □

§6. Case of curves

In addition to the hypotheses of §§4 and 5, we assume that V is an *algebraic curve* $(r = 1)$ and we denote by X the unique non-singular, complete model of the field $K = k(V)$.

28. Comparison of the divisor class group and generalized Jacobians

Let $\mathfrak{m}$ be a modulus on X, rational over k (cf. chap. V) and let $J_\mathfrak{m}$ be the corresponding generalized Jacobian. We know (*loc. cit.*, no. 22) that $J_\mathfrak{m}$ is defined over k and that X is equipped with a canonical map

$$\varphi_\mathfrak{m} : X \to H_\mathfrak{m}$$

to a principal homogeneous space $H_\mathfrak{m}$ for the group $J_\mathfrak{m}$ (we denote this homogeneous space by $H_\mathfrak{m}$ instead of $J_\mathfrak{m}^{(1)}$ to harmonize the notations with those of §3).

Proposition 21.
i) *The map $\varphi_\mathfrak{m} : X \to H_\mathfrak{m}$ is an element of L_k, in other words it is a maximal map defined over k.*
ii) *As $\mathfrak{m}$ runs through the set of moduli on X which are rational over k, the $\varphi_\mathfrak{m}$ form a cofinal system in L_k.*

PROOF. Let $H' \to H_\mathfrak{m}$ be an isogeny and suppose that $\varphi_\mathfrak{m}$ lifts to a rational map $\psi : X \to H'$. Because $\varphi_\mathfrak{m}$ is regular away from the support of $\mathfrak{m}$, the same is true of ψ (chap. III, prop. 14). Thus ψ factors as $\theta \circ \varphi_\mathfrak{m}$ where $\theta : H_\mathfrak{m} \to H'$ is a morphism. The composition $H_\mathfrak{m} \xrightarrow{\theta} H' \to H_\mathfrak{m}$ is the identity on $\varphi_\mathfrak{m}(X)$, thus everywhere, and this shows that the isogeny $H' \to H_\mathfrak{m}$ is trivial. As $\varphi_\mathfrak{m}$ is defined over k, assertion i) is proved.

On the other hand, let $\alpha : X \to H_\alpha$ be an arbitrary element of L_k. We know that α admits a modulus $\mathfrak{m}$. After replacing $\mathfrak{m}$ by the sum of its conjugates, we can assume that $\mathfrak{m}$ is rational over k. Proposition 13 of chap. V (or theorem 2 of chap. V, combined with lemma 6 of §3) shows that $\varphi_\mathfrak{m} \geq \alpha$, whence assertion ii). $\qquad\square$

We keep the same notations and denote by $C_\mathfrak{m}(k)$ the group of *divisor classes rational over k, modulo $\mathfrak{m}$-equivalence*. A divisor D, rational over k and disjoint from the support S of $\mathfrak{m}$, is considered $\mathfrak{m}$-equivalent to 0 if it is of the form $D = (g)$, with $g \in K^*$ and $g \equiv 1 \bmod \mathfrak{m}$. For every extension k'/k one can define in the same way the group $C_\mathfrak{m}(k')$. The Galois group $\mathfrak{g}$ of k'/k acts naturally on $C_\mathfrak{m}(k')$.

Lemma 13. *The canonical map from $C_\mathfrak{m}(k)$ to $C_\mathfrak{m}(k')$ is injective and its image is the set of elements of $C_\mathfrak{m}(k')$ invariant by $\mathfrak{g}$.*

PROOF. Let $K_\mathfrak{m}^*$ be the subgroup of K^* formed by the elements $g \equiv 1 \bmod \mathfrak{m}$, and let $D_\mathfrak{m}$ be the group of divisors rational over k and prime to S. Let $K_\mathfrak{m}'^*$ and $D_\mathfrak{m}'$ be the corresponding groups for k'. If we assume $\mathfrak{m} \neq 0$, the relation $(g) = 0$ implies $g = 1$ if $g \in K_\mathfrak{m}'^*$. Thus we have an exact sequence of $\mathfrak{g}$-modules

$$0 \to K_\mathfrak{m}'^* \to D_\mathfrak{m}' \to C_\mathfrak{m}(k') \to 0. \qquad (*)$$

It is clear that $H^0(\mathfrak{g}, K_\mathfrak{m}') = K_\mathfrak{m}^*$ and $H^0(\mathfrak{g}, D_\mathfrak{m}') = D_\mathfrak{m}$. The exact sequence of cohomology associated to $(*)$ can thus be written

$$0 \to K_\mathfrak{m}^* \to D_\mathfrak{m} \to H^0(\mathfrak{g}, C_\mathfrak{m}(k')) \to H^1(\mathfrak{g}, K_\mathfrak{m}'^*)$$

or as

$$0 \to C_m(k) \to H^0(\mathfrak{g}, C_m(k')) \to H^1(\mathfrak{g}, K_m'^*).$$

To prove lemma 13, it suffices to prove that $H^1(\mathfrak{g}, K_m'^*) = 0$, a result analogous to the classic "theorem 90". We are going to translate the proof of that theorem: choose $a \in k'$ such that $\mathrm{Tr}_{k'/k}(a) = 1$; if f_σ is a 1-cocycle with values in $K_m'^*$, we put

$$g = \sum_{\sigma \in \mathfrak{g}} a^\sigma f_\sigma.$$

As one can write $g = 1 + \sum a^\sigma (f_\sigma - 1)$, we have $g \equiv 1 \bmod m$. On the other hand, a direct computation shows that $f = g/g^\sigma$ and we have indeed proved that $H^1(\mathfrak{g}, K_m'^*) = 0$.

In the case $m = 0$, we have $K_m^* = K^*$, and the kernel of the map $K'^* \to D'$ is the group k'^*. One then modifies the preceding argument slightly, using the fact that $H^1(\mathfrak{g}, k'^*) = H^2(\mathfrak{g}, k'^*) = 0$ since k' is a finite field. $\qquad\square$

If D is a divisor prime to S, its image by φ_m is a well defined element of the group $H_m(\overline{k})$ attached to the homogeneous space H_m.

Proposition 22. *The map φ_m defines an isomorphism from $C_m(k)$ to the group $H_m(k)$.*

PROOF. We know (chap. V, thm. 1) that φ_m defines an isomorphism from $C_m(\overline{k})$ to $H_m(\overline{k})$. If we denote by G the Galois group of $\overline{k}/k$, lemma 13 shows that $C_m(k)$ is identified with the subgroup of $C_m(\overline{k})$ formed by the elements fixed by G. On the other hand, it is trivial that the same property holds for $H_m(k)$, whence the result. $\qquad\square$

Corollary. *Theorems 3, 3' and 3" of no. 25 are true for an algebraic curve.*

PROOF. We prove theorem 3'. Since the φ_m are cofinal in L_k, we are reduced to showing that, for every model V of K, the map

$$\varphi_m : Z_k(V) \to H_m(k)$$

is surjective. As V is biregularly isomorphic to X outside a finite set, we must prove that, if S' is a finite subset of X containing S, every element $x \in H_m(k)$ is the image by φ_m of a divisor rational over k and prime to S. According to proposition 22, we have in any case $x = \varphi_m(D)$ where D is prime to S. According to the approximation theorem, it is possible to find a function $g \in K^*$, $g \equiv 1 \bmod m$ and $v_P(g) = v_P(D)$ for all $p \in S' - S$. The divisor $D' = D - (g)$ answers the question. $\qquad\square$

[*Variant*: We choose a modulus $\mathfrak{m}'$ rational over k with support containing S', with $\mathfrak{m}' \geq \mathfrak{m}$. According to no. 16, the element $x \in H_{\mathfrak{m}}(k)$ is the image of an element $x' \in H_{\mathfrak{m}'}(k)$, which, by virtue of proposition 22, is the image of a divisor D' rational over k and prime to S'; *a fortiori* we have $\varphi_{\mathfrak{m}}(D') = x$.]

29. The idèle class group

We first recall the definition of this group:

Let I be the group of $\bar{k}$-idèles of X, i.e., the group of systems $(g_P)_{P \in X}$, $g_P \in \hat{L}_P$, $v_P(g_P) = 0$ for almost all $P \in X$ (we have written L for the field $K\bar{k}$ and $\hat{L}_P$ for its completion for the topology defined by the valuation v_P). The group G of k-automorphisms of $\bar{k}$ acts on I and the invariants of this group form the group $I(k)$ of k-*idèles*. In order that (g_P) belong to $I(k)$ it is necessary and sufficient that $g_P \in \hat{K}_P$ and $g_P = g_{P'}$ if P and P' are conjugate over k. The group K^* is identified with a subgroup of $I(k)$ and the quotient is *the group $C(k)$ of idèle classes* (over k).

If $\mathfrak{m} = \sum n_P P$ is a modulus rational over k, we denote by $I_{\mathfrak{m}}(k)$ the subgroup of $I(k)$ formed by idèles (g_P) such that $v_P(1 - g_P) \geq n_P$ if $P \in \mathrm{Supp}(\mathfrak{m})$ and $v_P(g_P) = 0$ if $P \notin \mathrm{Supp}(\mathfrak{m})$. Thus we get a decreasing filtered family of subgroups of $I(k)$ and one easily checks that $C(k)$ is identified with the projective limit of the groups $I(k)/K^* . I_{\mathfrak{m}}(k)$. On the other hand, $I(k)/K^* . I_{\mathfrak{m}}(k) = C_{\mathfrak{m}}(k)$ and proposition 22 permits us to identify $C_{\mathfrak{m}}(k)$ with $H_{\mathfrak{m}}(k)$. As the $\varphi_{\mathfrak{m}}$ form a cofinal subset of L_k, the projective limit of the $H_{\mathfrak{m}}(k)$ is equal to the group $A(K)$ of cycle classes of X (cf. no. 16). In other words:

Proposition 23. *The group $C(k)$ of idèle classes of X is canonically isomorphic to the group $A(K)$ of cycle classes of X.*

By theorem 1 of no. 16 we recover the fundamental theorem of class field theory for a function field in one variable:

Theorem 4. *The group of idèle classes of X is canonically isomorphic to the (modified) Galois group of the maximal Abelian extension of K.*

Of course, we also get the fact that this isomorphism is given by the reciprocity map.

One can also recover the results about the *conductor* of a finite Abelian extension L/K. If $\mathfrak{g}$ denotes its Galois group, we have $\mathfrak{g} = C(k)/N = I(K)/N'$, where N' is an open subgroup of $I(k)$ containing K^*; the conductor (in the arithmetic sense) of L/K is then the smallest modulus $\mathfrak{m}$ such that $I_{\mathfrak{m}}(k) \subset N'$; it comes to the same to say that it is the smallest $\mathfrak{m}$ such that $\mathfrak{g}$ is a quotient of $C_{\mathfrak{m}}(k)$. As $C_{\mathfrak{m}}(k) = H_{\mathfrak{m}}(k)$ is the Galois group of the field $E_{\mathfrak{m}}$ defined by the maximal map $\varphi_{\mathfrak{m}}$, we see that the conductor in question is the smallest $\mathfrak{m}$ such that L/K is "of type $\varphi_{\mathfrak{m}}$" in the sense of no. 19. Comparing with prop. 11, we finally get:

Proposition 24. *The conductor (in the arithmetic sense) of a finite Abelian extension L/K coincides with the conductor (in the geometric sense) of the extension $L\overline{k}/K\overline{k}$. Its support is the set of ramification points of L/K.*

30. Explicit reciprocity laws

Let L/K be a finite Abelian extension with Galois group $\mathfrak{g}$. By virtue of theorem 4, we have a surjective homomorphism $C(k) \to \mathfrak{g}$, whence a continuous homomorphism $I(k) \to \mathfrak{g}$. If $\mathfrak{p}$ is a rational prime cycle on X, and if g is an element of the field $\widehat{K}_{\mathfrak{p}}$, the completion of the field K for the topology defined by the valuation ring $\mathcal{O}_{\mathfrak{p}}$, we can consider the idèle g whose component at P is equal to g if $P \in \mathfrak{p}$ and equal to 1 if $P \notin \mathfrak{p}$. Its image in $\mathfrak{g}$ will be denoted $(L/K, g)_P$. As every idèle is congruent modulo $I_{\mathfrak{m}}(k)$ to a product of idèles of this type, *knowledge of the local symbols $(L/K, g)_P$ determines the homomorphism from the group $C(k)$ of idèle classes to the Galois group $\mathfrak{g}$.* If $\mathfrak{m}$ denotes the conductor of L/K and if S is its support, we have:

i) $(L/K, gg')_{\mathfrak{p}} = (L/K, g)_{\mathfrak{p}} + (L/K, g')_{\mathfrak{p}}$.

ii) $(L/K, g)_{\mathfrak{p}} = 0$ if $\mathfrak{p} \subset S$ and if $g \equiv 1 \bmod \mathfrak{m}$ at $\mathfrak{p}$.

iii) $(L/K, g)_{\mathfrak{p}} = v_{\mathfrak{p}}(g)(\mathfrak{p}, L/K)$ if $\mathfrak{p} \subset X - S$.

iv) $\sum_{\mathfrak{p}}(L/K, g)_{\mathfrak{p}} = 0$ for all $g \in K^*$.

Property i) expresses the fact that $I(k) \to \mathfrak{g}$ is a homomorphism. Properties ii) and iii) express the fact that this homomorphism is zero on $I_{\mathfrak{m}}(k)$, and defines by passage to the quotient a homomorphism from the group of divisors prime to S to $\mathfrak{g}$ which is nothing other than the reciprocity map. Finally, property iv) expresses the fact that $I(k) \to \mathfrak{g}$ is zero on K^*.

Thus we see that $(L/K, g)_{\mathfrak{p}}$ plays the role of a *local symbol* (in the sense of chap. III, §1) with respect to the Frobenius substitution $(\mathfrak{p}, L/K)$.

We make this more precise in a particular case:

Suppose that L/K is the pull-back of an isogeny $G' \to G$ by a rational map $f : X \to G$ (of course we suppose that G, G' and f are defined over k and that the isogeny $G' \to G$ is Abelian over k). According to prop. 6 of §1, the isogeny $G' \to G$ is a quotient of the isogeny $\wp : G \to G$. Let π be the canonical homomorphism from G_k to $\mathfrak{g}$.

Proposition 26. *For every rational prime cycle* $\mathfrak{p}$ *and every function* $g \in K^*$, *we have*

$$(L/K, g)_{\mathfrak{p}} = \pi \left(\sum_{P \in \mathfrak{p}} (f, g)_P \right). \tag{*}$$

(One can also take g in $\widehat{K}_{\mathfrak{p}}^*$, which changes nothing, in view of property iii) of local symbols.)

PROOF. Let $S' \supset S$ be the set of points where f is not regular. If $\mathfrak{p}$ is prime to S', the reasoning of no. 24 shows that

$$(\mathfrak{p}, L/K) = f(\mathfrak{p}) = \sum_{P \in \mathfrak{p}} f(P),$$

whence the formula (*) in this case.

If $\mathfrak{p} \subset S'$, we choose an auxiliary function $g_1 \in K^*$ approximating g at the points of $\mathfrak{p}$ and approximating 1 at the points of $S' - \mathfrak{p}$. Applying the result we have just proved to this function and using formula iv), we get the desired relation.   □

Examples. 1) Artin-Schreier extensions. Suppose that L/K is cyclic of degree p and choose a generator of the Galois group, thus identifying $\mathbf{F}_p$ with $\mathbf{Z}/p\mathbf{Z}$. The extension L/K is obtained as the pull-back of the isogeny $x \to x^p - x$ of $\mathbf{G}_a$ by a rational map $f : X \to \mathbf{G}_a$. We propose to compute explicitly the local symbol $(L/K, g)_{\mathfrak{p}}$ which one also denotes $[f, g]_{\mathfrak{p}}$. We apply prop. 26; the homomorphism $\pi : G_k \to \mathfrak{g}$ is here identified with the trace operation $\mathrm{Tr}_{k/\mathbf{F}_p} : k \to \mathbf{F}_p$. On the other hand, we know (chap. III, prop. 5) that $(f, g)_P = \mathrm{Res}_P(f dg/g)$ which is an element of the field $k(P)$. We deduce that

$$[f, g]_{\mathfrak{p}} = \mathrm{Tr}_{k/\mathbf{F}_p} \left(\sum_{P \in \mathfrak{p}} \mathrm{Res}_P(f dg/g) \right) = \mathrm{Tr}_{k/\mathbf{F}_p}(\mathrm{Tr}_{k(P)/k} \mathrm{Res}_P(f \, dg/g)),$$

whence finally

$$[f, g]_{\mathfrak{p}} = \mathrm{Tr}_{k(P)/\mathbf{F}_p}(\mathrm{Res}_P(f dg/g)), \qquad \text{with } P \in \mathfrak{p}.$$

2) Kummer extensions. Suppose that L/K is cyclic of degree n prime to the characteristic and that k contains a primitive n-th root of unity (which amounts to saying that q is divisible by $n-1$). Then we can identify the Galois group $\mathfrak{g}$ with the group of n-th roots of unity. The extension L/K is the pull-back of the isogeny $x \to x^n$ of $\mathbf{G}_m$ by the rational map $f : X \to \mathbf{G}_m$. The corresponding symbol $(L/K, g)_{\mathfrak{p}}$ is denoted $(f, g)_{\mathfrak{p}}$; it is Hilbert's *norm residue symbol*. One computes it by means of prop. 26. The homomorphism $\pi : G_k \to \mathfrak{g}$ is identified with the map $x \to x^{(q-1)/n}$

from k^* to the group of n-th roots of unity. As for the local symbol $(f,g)_\mathfrak{p}$, we know its value (chap. III, prop. 6). Thus we get

$$(f,g)_\mathfrak{p} = \left(\mathrm{N}_{k(P)/k}((-1)^{\alpha\beta} \frac{f^\beta}{g^\alpha}(P))) \right)^{\frac{q-1}{n}} \quad \text{with} \quad \begin{cases} \alpha = v_\mathfrak{p}(f) \\ \beta = v_\mathfrak{p}(g) \end{cases}.$$

These formulas are due to H. L. Schmidt [72].

§7. Cohomology

One knows that class field theory has been enriched by cohomological properties that can be summarized in the following statement:

(*) *The map which associates to any field* (of numbers or of functions over a finite field) *the corresponding idèle class group is a "class formation" in the sense of Artin-Tate.*

One can ask if the "cycle class group" that we have defined in §4 also satisfies (*). This is so in dimension 1, as is well known; we will see that this result is an immediate consequence of theorem 4 of §6. On the other hand, in dimension > 1, we will see that the cycle class group *is not* a class formation.

31. A criterion for class formations

We return to the situation of §4 and let $K = k(V)$ be an algebraic function field over a finite field k. Let E and F be two finite extensions of K, with F Galois over E with Galois group $\mathfrak{g}_{F/E}$. Let F_a (resp. E_a) be the maximal Abelian extension of F (resp. E) and let $A(F)$ (resp. $A(E)$) be its modified Galois group as explained in no. 16.

The extension F_a/E is Galois and contains the extension $E\overline{k}/k$. We denote by $\widehat{G}_{F/E}$ its Galois group (in the usual sense) and by $G_{F/E}$ its Galois group modified by the procedure of no. 16. By definition, $G_{F/E}$ is the subgroup of $\widehat{G}_{F/E}$ formed by the elements σ such that there exists $n \in \mathbf{Z}$ with $\sigma(a) = a^{q^n}$ for $a \in \overline{k}$. The group $G_{F/E}$ is topologized so that the Galois group of $F_a/E\overline{k}$, call it $G^0_{F/E}$, is an open subgroup. This group has properties very close to those of a Galois group; its closed subgroups correspond to fields L, with $F_a \supset L \supset E$, such that $L \cap \overline{k}$ is equal either to $\overline{k}$ or to a finite extension of k. The group $\widehat{G}_{F/E}$ is just the completion of $G_{F/E}$ for the topology defined by the closed subgroups of finite index.

The group $G_{F/E}$ is an extension of $A(F)$ by $\mathfrak{g}_{F/E}$; this extension corresponds to a cohomology class $u_{F/E} \in H^2(\mathfrak{g}_{F/E}, A(F))$. The statement (*) can then be made more precise as follows:

Definition 6. One says that the map $E \to A(E)$ is a class formation if, for every pair F/E, with $F \supset E \supset K$, with F finite over K and F Galois over E,

i) $H^1(\mathfrak{g}_{F/E}, A(F)) = 0$.

ii) $H^2(\mathfrak{g}_{F/E}, A(F))$ is cyclic of order $[F : E]$ and generated by $u_{F/E}$.

Remark. Let $\widehat{A}(F)$ be the Galois group (in the usual sense) of F_a/F. The quotient group $\widehat{A}(F)/A(F)$ is isomorphic to $\widehat{\mathbf{Z}}/\mathbf{Z}$, which is a *divisible group without torsion*. One deduces from this that $E \to \widehat{A}(E)$ *is a class formation if and only if* $E \to A(E)$ *is one.* Furthermore, as $\widehat{A}(F)$ is a compact group, the same is true of $H^q(\mathfrak{g}_{F/E}, \widehat{A}(F))$, and the $H^q(\mathfrak{g}_{F/E}, A(F))$, with the topology deduced from that of $A(F)$, are *separated* groups.

We return now to the extension

$$1 \to A(F) \to G_{F/E} \to \mathfrak{g}_{F/E} \to 1.$$

The group $A(F)$ is of finite index in $G_{F/E}$. But one knows that if H is an Abelian subgroup of finite index of a group G, then one can define a homomorphism $T : G/G' \to H$ called the *transfer* (G' denoting the group of commutators of G). We recall the definition of this homomorphism (for more details, see Zassenhaus [104], chap. V and Cartan-Eilenberg [10], chap. XII): if G/H denotes the homogeneous space of right cosets of H in G, one chooses a section $s : G/H \to G$; if $g \in G$ and $x \in G/H$, one has $s(xg) \equiv s(x).g \bmod H$, so there exists an element $h_{g,x} \in H$ such that $s(x).g = h_{g,x}.s(xg)$. One then puts

$$T(g) = \prod_{x \in G/H} h_{g,x}$$

and one checks by a direct computation that this is a homomorphism from the Abelianization of G to H and that this homomorphism does not depend on the choice of the section s.

Thus we have a homomorphism $T : G_{F/E}/G'_{F/E} \to A(F)$. This homomorphism being continuous, its kernel contains the closure $G^c_{F/E}$ of $G'_{F/E}$. But, since E_a is the largest Abelian extension of E contained in F_a, the subgroup $G^c_{F/E}$ is the closed subgroup of $G_{F/E}$ corresponding to E_a and $A(E) = G_{F/E}/G^c_{F/E}$. Thus we have obtained a continuous homomorphism

$$\mathrm{Ver} : A(E) \to A(F).$$

The image of this homorphism is contained in $H^0(\mathfrak{g}_{F/E}, A(F))$, the subset of $A(F)$ formed by the elements invariant by $\mathfrak{g}_{F/E}$. Furthermore, the composition

$$A(F) \to A(E) \to A(F)$$

is nothing other than the *trace* (the first homomorphism being that defined in no. 21).

Theorem 5. *In order that $E \to A(E)$ be a class formation, it is necessary and sufficient that, for every pair F/E satisfying the conditions of definition 6, the homomorphism*

$$\text{Ver} : A(E) \to A(F)$$

be injective and have for image $H^0(\mathfrak{g}_{F/E}, A(F))$.

This theorem is implicit in the memoirs of Weil [91] and of Hochschild-Nakayama [36]; it was made explicit by Artin-Tate (non-published). The fact that the condition is *necessary* can also be found in Kawada [41]; we are going to rapidly sketch a proof.

Suppose that $A(E)$ is a class formation. If we denote by $\widehat{H}^q(\mathfrak{g})$ the *modified* cohomology groups of $\mathfrak{g}$ (see Cartan-Eilenberg [10], chap. XII), the theorem of Tate [83] shows that

$$\widehat{H}^q(\mathfrak{g}_{F/E}, A(F)) = \widehat{H}^{q-2}(\mathfrak{g}_{F/E}, \mathbf{Z}) \quad \text{for all } q \in \mathbf{Z}.$$

Write $\mathfrak{g} = \mathfrak{g}_{F/E}$ to simplify the notation. In particular

$$\widehat{H}^{-1}(\mathfrak{g}, A(F)) = \widehat{H}^{-3}(\mathfrak{g}, \mathbf{Z}) = H_2(\mathfrak{g}, \mathbf{Z})$$

and

$$\widehat{H}^0(\mathfrak{g}, A(F)) = \widehat{H}^{-2}(\mathfrak{g}, \mathbf{Z}) = H_1(\mathfrak{g}, \mathbf{Z}) = \mathfrak{g}/\mathfrak{g}'.$$

But, by the definition of the modified groups $\widehat{H}^{-1}$ and $\widehat{H}^0$, there is an exact sequence

$$0 \to \widehat{H}^{-1}(\mathfrak{g}, A(F)) \to H_0(\mathfrak{g}, A(F)) \to H^0(\mathfrak{g}, A(F)) \to \widehat{H}^0(\mathfrak{g}, A(F)) \to 0.$$

This exact sequence here takes the form

$$H_2(\mathfrak{g}, \mathbf{Z}) \to H_0(\mathfrak{g}, A(F)) \to H^0(\mathfrak{g}, A(F)) \to \mathfrak{g}/\mathfrak{g}' \to 0. \qquad (**)$$

On the other hand, the exact sequence of homology associated to the extension $G_{F/E}/A(F) = \mathfrak{g}$ gives (cf. Cartan-Eilenberg [10], p. 350)

$$H_2(\mathfrak{g}, \mathbf{Z}) \to H_0(\mathfrak{g}, A(F)) \to G_{F/E}/G'_{F/E} \to \mathfrak{g}/\mathfrak{g}' \to 0. \qquad (***)$$

From this last sequence and from the fact that $H_0(\mathfrak{g}, A(F))$ is a *separated* topological group (cf. above), we deduce that $G'_{F/E}$ is *closed* in $G_{F/E}$, whence the fact that

$$G_{F/E}/G'_{F/E} = A(E).$$

Then comparing the exact sequences (**) and (***) and showing that the transfer Ver : $A(E) \to H^0(\mathfrak{g}, A(F))$ satisfies the necessary compatibility conditions, one deduces that the transfer is bijective, which proves the first part of the theorem.

It remains to prove that the condition of theorem 5 is *sufficient*; This is what we will do in the two nos. that follow.

32. Some properties of the cohomology class $u_{F/E}$

Lemma 14. *If $F \supset F' \supset E$, with F finite and Galois over E, the image of $u_{F/E}$ in $H^2(\mathfrak{g}_{F/F'}, A(F))$ is $u_{F/F'}$.*

(The homomorphism in question is the one

$$H^2(\mathfrak{g}_{F/E}, A(F)) \to H^2(\mathfrak{g}_{F/F'}, A(F))$$

induced by the injection $\mathfrak{g}_{F/F'} \to \mathfrak{g}_{F/E}$.)

PROOF. One has a "tower" of fields $E \subset F' \subset F \subset F_a$, and thus a commutative diagram

$$
\begin{array}{ccccccccc}
1 & \longrightarrow & A(F) & \longrightarrow & G_{F/F'} & \longrightarrow & \mathfrak{g}_{F/F'} & \longrightarrow & 1 \\
& & \downarrow & & \downarrow & & \downarrow & & \\
1 & \longrightarrow & A(F) & \longrightarrow & G_{F/E} & \longrightarrow & \mathfrak{g}_{F/E} & \longrightarrow & 1.
\end{array}
$$

The lemma follows immediately from this.   □

Lemma 15. *Let $\mathfrak{g}$ be a group, $\mathfrak{h}$ a normal subgroup of finite order and let $\sigma = \sum_{x \in \mathfrak{h}} x$, considered as an element of the group algebra $\mathbf{Z}(\mathfrak{g})$ of the group $\mathfrak{g}$. If M is a $\mathfrak{g}$-module, the endomorphism of $H^q(\mathfrak{g}, M)$ defined by $\sigma : M \to M$ is multiplication by n.*

PROOF. Because $\mathfrak{h}$ is normal in $\mathfrak{g}$, the element σ is a sum of classes, thus belongs to the *center* of $\mathbf{Z}(\mathfrak{g})$; further, the image of σ by the "augmentation" homomorphism $\mathbf{Z}(\mathfrak{g}) \to \mathbf{Z}$ is n. The lemma follows from that and from the formula defining the cohomology groups ([10], chap. X)

$$H^q(\mathfrak{g}, M) = Ext^q_{\mathbf{Z}(\mathfrak{g})}(\mathbf{Z}, M).$$ □

[This is the same reasoning that, in the theory of Lie algebras, shows that the Casimir operator acts trivially on cohomology.]

Lemma 16. *If $F \supset F' \supset E$, with F and F' Galois over E, the image u' of $u_{F'/E}$ in $H^2(\mathfrak{g}_{F/E}, A(F))$ is equal to $[F : F']u_{F/E}$.*

(The homomorphism is $H^2(\mathfrak{g}_{F'/E}, A(F)) \to H^2(\mathfrak{g}_{F/E}, A(F))$ induced by $\mathfrak{g}_{F/E} \to \mathfrak{g}_{F'/E}$ and $Ver : A(F') \to A(F)$.)

PROOF. Here again, there is a commutative diagram

$$
\begin{array}{ccccccccc}
1 & \longrightarrow & A(F) & \longrightarrow & G_{F/E} & \longrightarrow & \mathfrak{g}_{F/E} & \longrightarrow & 1 \\
 & & \downarrow & & \downarrow & & \downarrow & & \\
1 & \longrightarrow & A(F') & \longrightarrow & G_{F'/E} & \longrightarrow & \mathfrak{g}_{F'/E} & \longrightarrow & 1.
\end{array}
$$

We deduce from this that $u_{F/E}$ and $u_{F'/E}$ have the same image u'' in the group $H^2(\mathfrak{g}_{F/E}, A(F'))$. But the image of u'' in $H^2(\mathfrak{g}_{F/E}, A(F))$ by Ver : $A(F') \to A(F)$ is equal to u'. It follows that u' is deduced from $u_{F/E}$ by the homomorphism $A(F) \to A(F') \to A(F)$, which itself is just the *trace* with respect to the action of $\mathfrak{g}_{F/F'}$. Applying lemma 15 to the normal subgroup $\mathfrak{g}_{F/F'}$ of $\mathfrak{g}_{F/E}$ we deduce that $u' = [F : F']u_{F/E}$. □

33. Proof of theorem 5

We must prove the *sufficiency* of the condition of theorem 5. Thus we suppose that, for every finite Galois extension F/E with E a finite extension of K, the transfer homomorphism Ver : $A(E) \to A(F)$ is injective and its image is the set of elements of $A(F)$ invariant by $\mathfrak{g}_{F/E}$.

Lemma 17. $H^1(\mathfrak{g}_{F/E}, A(F)) = 0$.

PROOF. The hypothesis on the transfer permits us to identify $A(E)$ with a subgroup of $A(F)$ and we will do this from now on. The usual reasoning of "dévissage" using Sylow subgroups (cf. for example Chevalley [16], or Hochschild-Nakayama [36]) shows that it suffices to prove the vanishing of $H^1(\mathfrak{g}_{F/E}, A(F))$ when $\mathfrak{g}_{F/E}$ is a cyclic group. Let σ be a generator of this group. There is an extension

$$1 \to A(F) \to G_{F/E} \to \mathfrak{g}_{F/E} \to 1$$

and the subgroup $G'_{F/E}$ of commutators of $G_{F/E}$ is the subgroup of $A(F)$ formed by the $b - \sigma(b)$, $b \in A(F)$. By hypothesis, the transfer Ver : $G_{F/E}/G'_{F/E} \to A(F)$ is injective. Thus if $a \in A(F)$ is an element with trace zero, we must have $a \in G'_{F/E}$, i.e., $a = b - \sigma(b)$, which is equivalent to the vanishing of $H^1(\mathfrak{g}_{F/E}, A(F))$. □

Lemma 18. *The element $u_{F/E}$ has order $[F : E]$.*

PROOF. One knows that the order of $u_{F/E}$ *divides* $[F : E]$; everything comes down to showing that it is not smaller.

First suppose that F/E is cyclic of prime order l. If $u_{F/E}$ is not of order l, necessarily $u_{F/E} = 0$, in other words $G_{F/E}$ is a *semi-direct product* of $\mathfrak{g}_{F/E}$ by $A(F)$. In this case, the transfer is not difficult to determine explicitly: it is zero on $\mathfrak{g}_{F/E}$ (identified with a subgroup of $G_{F/E}$) and it is the trace on $A(F)$. On the other hand, since $\mathfrak{g}_{F/E}$ is cyclic, we have seen that the commutator subgroup of $G_{F/E}$ is contained in $A(F)$ and, as we have seen, it does not contain $\mathfrak{g}_{F/E}$. The transfer

$$\text{Ver} : G_{F/E}/G'_{F/E} \to A(F)$$

thus has a non-trivial kernel, contrary to the hypothesis.

Now we suppose that F/E is of order l^n and we argue by induction on n. If $u_{F/E}$ is not of order l^n, then $l^{n-1} u_{F/E} = 0$. According to the properties of p-groups, there exists a subfield F' of F, cyclic of degree l over E. Thus $[F : F'] = l^{n-1}$ and lemma 16 shows that the image of $u_{F'/E}$ in $H^2(\mathfrak{g}_{F/E}, A(F))$ is zero. But lemma 17, together with a well-known exact sequence, shows that the homomorphism $H^2(\mathfrak{g}_{F'/E}, A(F')) \to H^2(\mathfrak{g}_{F/E}, A(F))$ is injective. Thus $u_{F'/E}$ is zero, contrary to what we have just seen.

The general case follows immediately from the above, using the Sylow subgroups of $\mathfrak{g}_{F/E}$ as well as lemma 14.   □

This lemma shows that $H^2(\mathfrak{g}_{F/E}, A(F))$ contains a cyclic subgroup of order $[F : E]$. To finish the proof of theorem 5, it thus suffices to prove the following result:

Lemma 19. *The order of the group $H^2(\mathfrak{g}_{F/E}, A(F))$ is $\leq [F : E]$.*

PROOF. By the same "dévissage" as before, we are reduced to the case of a cyclic extension of prime order l. Let B be the subset of $A(F)$ formed by the elements invariant by $\mathfrak{g}_{F/E}$, and let T be the subset of B which is the image of the map $\text{Tr} : A(F) \to A(F)$. Then $H^2(\mathfrak{g}_{F/E}, A(F)) = B/T$ and thus we must prove that $(B : T) \leq l$. But, by hypothesis, the homomorphism $\text{Ver} : A(E) \to B$ is surjective, and composing it with $A(F) \to A(E)$ we find the trace. It follows that B/T is the image of the cokernel of $A(F) \to A(E)$, which is cyclic of order l, whence the result.   □

34. Map to the cycle class group

Let r be the dimension of V, i.e., the transcendence degree of the extension K/k.

Theorem 6. *In order that $A(E)$ be a class formation, it is necessary and sufficient that $r \leq 1$.*

PROOF. In the case where $r = 0$, one has $A(E) = \mathbf{Z}$, and it is trivial that we get a class formation.

When $r = 1$, theorem 4 of §6 shows that $A(E) = C_E$, the idèle class group of the field E. The homomorphism Ver : $A(E) \to A(F)$ becomes a homomorphism

$$\text{Ver} : C_E \to C_F.$$

A simple computation with Frobeniuis substitutions (that by which Artin proved for the first time the "Hauptidealsatz" cf. for example [30]) shows that Ver : $C_E \to C_F$ is none other than the *canonical injection* of the first group in the second. Theorem 90 then shows that every element of C_F invariant by $\mathfrak{g}_{F/E}$ belongs to C_E (cf. Weil [91], §1), whence the fact that $A(E)$ is a class formation by virtue of theorem 5.

(Thus, in the case of function fields, the cohomological theory contains *nothing more* than the classical theory. The situation is different for number fields, where the presence of the connected component of the ideal class group prevents one from easily recovering the Galois group from the idèle class group, cf. Weil [91].)

We now show that, for $r \geq 2$, $A(E)$ *is not* a class formation. Suppose the contrary. Put $K_1 = K\overline{k}$ and denote by $A(K_1)$ the Galois group of the maximal Abelian extension of K_1. Similarly, define $A(E_1)$ for every finite extension E_1 of K_1. We are going to see that $A(E_1)$ is a *class formation*. Indeed let F_1/E_1 be a finite Galois extension with Galois group $\mathfrak{g}$. We can find a finite extension k'/k and extensions $F' \supset E' \supset Kk'$, with $F_1 = F'\overline{k}$, $E_1 = E'\overline{k}$, and F_1/E_1 Galois with group $\mathfrak{g}$. We deduce that $A(F_1)$ is equal to the projective limit $\varprojlim A(F'k'')$, over finite extensions k'' of k'. Furthermore, the diagram

$$
\begin{array}{ccc}
A(E') & \longrightarrow & A(E'k'') \\
\text{Ver} \downarrow & & \text{Ver} \downarrow \\
A(F') & \longrightarrow & A(F'k'')
\end{array}
$$

is commutative. Using the fact that the functor "projective limit" is left exact, we deduce that Ver : $A(E_1) \to H^0(\mathfrak{g}, A(F_1))$ is bijective, whence our assertion.

The field $K_1 = K\overline{k}$ contains all the roots of unityand thus is amenable to *Kummer theory*. More precisely, decomposing $A(K_1)$ into its p-primary component $A_p(K_1)$ and its prime-to-p component $A_*(K_1)$, the dual of the compact group $A_*(K_1)$ is nothing other than the projective limit $B(K_1)$ of the groups $B_n(K_1) = K_1^*/K_1^{*n}$, the limit being taken over the set of integers n prime to p (the homomorphism $K^*/K^{*n} \to K^*/K^{*nm}$ being the raising to the m-th power). By transport, the transfer becomes a homomorphism $N : B(F_1) \to B(E_1)$ which is the *norm*. Expressing that the image of Ver is equal to the set of elements of $A(F_1)$ left fixed by $\mathfrak{g}$, we find that, if $a \in F_1$ is such that $N\,a \in E_1^{*n}$, there exists m prime to p such that a^m is congruent mod F_1^{*mn} to a product of elements of the form $b^{1-\sigma}$, $\sigma \in \mathfrak{g}$, $b \in F_1^*$. Suppose in particular that $[F_1 : E_1] = n$ is prime to p, and take $a \in E_1^*$; then $N a = a^n$ and the preceding condition will be satisfied.

We will thus have

$$a^m = c^{mn} \prod b^{1-\sigma} \qquad b, c \in F_1^*.$$

Taking the norm of the two sides, we deduce

$$a^{mn} = (N c)^{nm},$$

whence

$$a = \epsilon . N c$$

where ϵ is a nm-th root of unity. But such a root is an n-th power in $\overline{k}$, thus a norm and we finally deduce that a *is a norm*, i.e., that *every element of E_1 is the norm of an element of F_1 if F_1/E_1 is Galois and of degree prime to p.*

This is true if $r = 1$ (for it is equivalent to saying that the Brauer group of K_1 contains no element of order prime to p, but one knows, by virtue of Tsen's theorem, that this group is trivial).

On the other hand, it is a contradiction of the hypothesis that $r \geq 2$. Indeed, this hypothesis implies, as one knows, that the field E_1 has a valuation v whose value group is $\mathbf{Z} \oplus \mathbf{Z}$. Let a be an element of E_1 such that $v(a) = (1, 0)$, and let F_1 be the field obtained by adjoining to E_1 the n-th root of a (n prime to p). We have $[F_1 : E_1] = n$ and the valuation v ramifies in F_1; if v_1 denotes one of its extensions, the value group of F_1 contains $\frac{1}{n}\mathbf{Z} \oplus \mathbf{Z}$. From this, and from the formula $\sum e_i f_i \leq n$, we deduce that v_1 is the unique extension of v, and that the value group of v_1 is exactly $\frac{1}{n}\mathbf{Z} \oplus \mathbf{Z}$. But every norm has a valuation which belongs to $n(v_1(F_1^*))$, thus to $\mathbf{Z} \oplus \mathbf{Z}$. In particular, an element of E_1 whose valuation is of the form $(0, 1)$ is not a norm, and this contradiction finishes the proof. $\qquad\square$

Bibliographic note

Class field theory for function fields of one variable over a finite field was first built up by F. K. Schmidt and Witt following the model of number fields, in other words by means of tedious index computations. Weil attempted as early as 1940 to recover these results by geometric means; this was shown in his note on the Riemann hypothesis [86], as well as in the last lines of his work on algebraic curves [88]. But the usual Jacobian, which he did have, was insufficient; generalized Jacobians, and for higher dimensional varieties the Albanese variety, were needed. Lang saw this ([48], [49], [50]), and most of the results and methods of this chapter are due to him. It was he who saw that class fields can be constructed *a priori* as pull-backs of isogenies of the type $\wp$, which makes the Artin reciprocity law evident. It is hardly necessary to say that one knows no analog for number fields (except in the case of the rationals and imaginary quadratic fields). Lang also developed an analytic theory (L-series) about which we have not spoken.

We mention a recent paper of Morikawa [57] which contains an exposé of the case of curves directly inspired by the first memoir of Lang [49]. Morikawa gives a direct proof (not using Rosenlicht's theorem of chap. III) of the fact that every Abelian covering of a curve is a pull-back of an isogeny of a generalized Jacobian.

Lang's method essentially comes down to considering the group of idèle classes of degree 0 as a proalgebraic group ([122], [123] §5). One can also apply this idea to local class field theory, both in equal and mixed characteristic, cf. [123] as well as Hazewinkel [112].

"Classical" class field theory is presented in Hasse [30], Herbrand [33], Lang [116], and Weil [125]. The cohomological theory, due to Weil [91], Hochschild-Nakayama [36], and Tate [83] can be found in Chevalley [16], Artin-Tate [106], and Cassels-Fröhlich [109]; see also Kawada [40], [41], [42] and Kawada-Satake [43].

CHAPTER VII

Group Extensions and Cohomology

In this chapter, we again assume that the base field k is *algebraically closed*.

All the algebraic groups considered are *commutative* (but not necessarily connected).

§1. Extensions of groups

Let $\mathcal{C}$ be the category of commutative groups (in the usual sense, that is not endowed with an algebraic group structure). If A, $B \in \mathcal{C}$, one knows how to define the group $\text{Ext}(A, B)$ of Abelian extensions of A by B, either by a direct procedure (Baer [2]) or by taking the first derived functor of the functor $\text{Hom}(A, B)$ (Cartan-Eilenberg [10], chap. XIV). One also knows that the elements of $\text{Ext}(A, B)$ can be interpreted as classes of symmetric factor systems.

The aim of this § is to transpose the definition of $\text{Ext}(A, B)$ and its principal properties to the category $\mathcal{C}_A$ of commutative algebraic groups.

1. The groups $\text{Ext}(A, B)$

Let A, B, C be three (commutative) algebraic groups. A sequence of (algebraic) homomorphisms

$$0 \to B \to C \to A \to 0 \tag{1}$$

is called *strictly exact* if it is exact in the usual sense, and if the homomorphisms $B \to C$ and $C \to A$ are separable, in other words if the algebraic structure of B (resp. of A) is induced by (resp. is the quotient of) that of C. We can then identify B with a subgroup of C and identify A with C/B.

We denote by t_A (resp. t_B, t_C) the tangent space to A (resp. to B, to C) at the origin. The sequence (1) gives rise to the sequence

$$0 \to t_B \to t_C \to t_A \to 0. \tag{2}$$

Lemma 1. *In order that the sequence* (1) *of algebraic groups be strictly exact, it is neccesary and sufficient that it be exact and that the sequence* (2) *of vector spaces also be exact.*

PROOF. This lemma is merely a reformulation of corollaries 2 and 3 to proposition 16 of chap. III. $\qquad\square$

A strictly exact sequence (1) is called an *extension* of A by B (in fact, we often abuse language by saying that the group C is an extension of A by B). Two extensions C and C' are *isomorphic* if there exists a homomorphism $f : C \to C'$ making a commutative diagram

$$
\begin{array}{ccccccccc}
0 & \longrightarrow & B & \longrightarrow & C & \longrightarrow & A & \longrightarrow & 0 \\
& & \downarrow{\scriptstyle id.} & & \downarrow{\scriptstyle f} & & \downarrow{\scriptstyle id.} & & \\
0 & \longrightarrow & B & \longrightarrow & C' & \longrightarrow & A & \longrightarrow & 0.
\end{array}
\tag{3}
$$

In this case, f is automatically an *isomorphism*. Indeed, it is clear that f is bijective and the diagram analogous to (3) formed by the tangent spaces t_A, t_B, t_C, and $t_{C'}$ shows that f defines an isomorphism from t_C to $t_{C'}$, so our assertion follows from lemma 1.

The set of *classes of extensions* of A by B (with respect to isomorphism) is denoted $\mathrm{Ext}(A, B)$. It is a *contravariant functor in A and a covariant functor in B*:

a) If $f : B \to B'$ is a homomorphism and if $C \in \mathrm{Ext}(A, B)$, one defines $f_*(C)$ (also denoted fC) as the quotient of $C \times B'$ by the subgroup formed by pairs $(-b, f(b))$ where b runs through B. The canonical maps $B' \to C \times B'$ and $C \times B' \to C$ define by passage to the quotient a sequence

$$0 \to B' \to f_*(C) \to A \to 0. \tag{4}$$

This sequence is strictly exact, as one can see by applying lemma 1; thus $f_*(C) \in \mathrm{Ext}(A, B')$.

One can also characterize $f_*(C)$ as the unique extension C' of A by B' such that there exists a homomorphism $F : C \to C'$ making a commutative diagram

$$
\begin{array}{ccccccccc}
0 & \longrightarrow & B & \longrightarrow & C & \longrightarrow & A & \longrightarrow & 0 \\
& & \downarrow{\scriptstyle f} & & \downarrow{\scriptstyle F} & & \downarrow{\scriptstyle id.} & & \\
0 & \longrightarrow & B' & \longrightarrow & C' & \longrightarrow & A & \longrightarrow & 0.
\end{array}
\tag{5}
$$

b) Similarly, if $g : A' \to A$ is a homomorphism and if $C \in \text{Ext}(A, B)$, one defines $g^*(C)$ (also denoted Cg) as the subgroup of $A' \times C$ formed by pairs (a', c) having the same image in A. It is an element of $\text{Ext}(A', B)$ that can be characterized by a diagram (6) analogous to (5) which the reader can make more explicit. More generally, let $C \in \text{Ext}(A, B)$, $C' \in \text{Ext}(A', B')$ and let $f : B \to B'$, $g : A \to A'$ be two homomorphisms. The relation

$$g^*(C') = f_*(C) \qquad \text{in} \quad \text{Ext}(A, B') \qquad (7)$$

is equivalent to the existence of a homorphism F making a commutative diagram

$$
\begin{array}{ccccccccc}
0 & \longrightarrow & B & \longrightarrow & C & \longrightarrow & A & \longrightarrow & 0 \\
 & & {\scriptstyle f}\downarrow & & {\scriptstyle F}\downarrow & & {\scriptstyle g}\downarrow & & \\
0 & \longrightarrow & B' & \longrightarrow & C' & \longrightarrow & A' & \longrightarrow & 0.
\end{array}
\qquad (8)
$$

It follows immediately that $\text{Ext}(A, B)$ is actually a *functor*, in other words that $1_* = 1$, $(f f')_* = f_* f'_*$, $1^* = 1$, and $(gg')^* = g'^* g^*$. Furthermore, $f_* g^* = g^* f_*$ (which justifies the notations fC and Cg).

We are now going to endow $\text{Ext}(A, B)$ with a composition law. We have only to adapt the method of Baer [2]: if C and C' are two elements of $\text{Ext}(A, B)$, the product $C \times C'$ can be considered as an element of $\text{Ext}(A \times A, B \times B)$. Denoting by respectively $d : A \to A \times A$ and $s : B \times B \to B$ the diagonal map of A and the composition law of B, we put

$$C + C' = d^* s_*(C \times C') \qquad \text{in} \quad \text{Ext}(A, B) \qquad (9)$$

and this is the desired composition law.

Proposition 1. *The composition law defined above makes $\text{Ext}(A, B)$ an Abelian group. If $\mathcal{C}_A$ denotes the additive category of commutative algebraic groups, the functor $\text{Ext}(A, B)$ is an additive bifunctor on $\mathcal{C}_A \times \mathcal{C}_A$.*

PROOF. The proof consists in checking that the constructions of Baer [2] applied to the present case yield algebraic groups (which is always immediate) and strictly exact sequences (by using lemma 1). This verification is somewhat tedious but presents no difficulties. Thus we limit ourselves to some brief indications:

a) The *neutral element*, denoted 0, of $\text{Ext}(A, B)$ is the *trivial extension* $C = A \times B$. Note that this extension is characterized by the existence of a section $A \to C$ (resp. a retraction $C \to B$) which is a homomorphism.

b) The *commutativity* of $\text{Ext}(A, B)$ is evident from the definition.

c) The *associativity* of $\text{Ext}(A, B)$ is proved by considering the sum map $s^3 : B \times B \times B \to B$ and the diagonal map $d_3 : A \to A \times A \times A$ and by showing that both of the compositions $(C + C') + C''$ and $C + (C' + C'')$ are equal to $d_3^* s_*^3 (C \times C' \times C'')$.

d) The *additivity* of the functor $\text{Ext}(A, B)$, that is to say the fact that $f_*(C)$ is bilinear in f and C (and similarly for $g^*(C)$) can be checked by constructing diagrams like (5) and (6).

e) The existence of the *inverse* of an element $C \in \text{Ext}(A, B)$ is shown by taking $(-1)^*(C)$, where -1 denotes the endomorphism $a \to -a$ of A. $\quad\square$

We make explicit two particular cases of proposition 1:

$$\begin{cases} \text{Ext}(A \times A', B) = \text{Ext}(A, B) \times \text{Ext}(A', B) \\ \text{Ext}(A, B \times B') = \text{Ext}(A, B) \times \text{Ext}(A, B'). \end{cases} \qquad (10)$$

2. The first exact sequence of Ext

If A and B are two algebraic groups, we denote by $\text{Hom}(A, B)$ the group of (algebraic) homomorphisms from A to B. We are going to see that $\text{Hom}(A, B)$ and $\text{Ext}(A, B)$ are related by an exact sequence analogous to that which holds in the category $\mathcal{C}$ of Abelian groups.

Let $0 \to A' \to A \to A'' \to 0$ be a strictly exact sequence; thus $A \in \text{Ext}(A'', A')$. If $\varphi \in \text{Hom}(A', B)$, then $\varphi_*(A) \in \text{Ext}(A'', B)$ and we put $d(\varphi) = \varphi_*(A)$. The map

$$d : \text{Hom}(A', B) \to \text{Ext}(A'', B)$$

thus defined is a homomorphism.

Proposition 2. *Let* $0 \to A' \to A \to A'' \to 0$ *be a strictly exact sequence and let B be an algebraic group. The sequence*

$$0 \to \text{Hom}(A'', B) \to \text{Hom}(A, B) \to \text{Hom}(A', B) \xrightarrow{d}$$
$$\text{Ext}(A'', B) \to \text{Ext}(A, B) \to \text{Ext}(A', B) \quad (11)$$

is exact.

(The homomorphisms figuring in this sequence are, with the exception of d, those defined canonically by $A' \to A$ and $A \to A''$.)

PROOF. We must check the exactness at $\text{Hom}(A'', B)$, ..., $\text{Ext}(A, B)$:

i) *Exactness at* $\text{Hom}(A'', B)$. This means that a homomorphism $A'' \to B$ is zero if and only if the composition $A \to A'' \to B$ is, which is clear since $A \to A''$ is surjective.

ii) *Exactness at* $\text{Hom}(A, B)$. This means that a homomorphism $A \to B$ induces 0 on A' if and only if it factors as $A \to A'' \to B$, which is clear since A'' is the quotient of A by A'.

iii) *Exactness at* $\text{Hom}(A', B)$. Let $\varphi \in \text{Hom}(A', B)$. We must see that φ can be extended to A if and only if the extension $C = \varphi_*(A) \in \text{Ext}(A'', B)$ is trivial. But, by definition, $C = A \times B / D_\varphi$, denoting by D_φ the subgroup of $A \times B$ formed by pairs $(-a', \varphi(a'))$, $a' \in A'$. If φ can be extended to A, this defines a homomorphism $\Phi : A \times B \to B$ which is null on D_φ, that is a homomorphism $C \to B$ which is a retraction and C is a trivial extension. Conversely, such a retraction defines Φ and thus an extension of φ to A.

iv) *Exactness at* $\text{Ext}(A'', B)$. Let $C \in \text{Ext}(A'', B)$. We must see that the image C_1 of C in $\text{Ext}(A, B)$ is trivial if and only if C is of the form $\varphi_*(A)$, with $\varphi \in \text{Hom}(A', B)$. But C_1 is trivial if and only there exists a section $A \to C_1$ which is a homomorphism, in other words if $A \to A''$ factors as $A \xrightarrow{\psi} C \to A''$. If $C = \varphi_*(A) = A \times B / D_\varphi$, such a factorisation is evident. Conversely, given such a factorisation, we define $\varphi : A' \to B$ as the restriction of ψ to A' and immediately check that the homomorphism $A \times B \to C$ defined by ψ identifies C with $A \times B / D_\varphi$, that is to say with $\varphi_*(A)$.

v) *Exactness at* $\text{Ext}(A, B)$. As the composition $A' \to A \to A''$ is zero, the same is true of $\text{Ext}(A'', B) \to \text{Ext}(A, B) \to \text{Ext}(A', B)$. We must show that , conversely, if $C_1 \in \text{Ext}(A, B)$ gives 0 in $\text{Ext}(A', B)$, the extension C_1 comes from an extension $C \in \text{Ext}(A'', B)$. But the hypothesis means that there exists a section homomorphism $A' \to C_1$. Putting $C = C_1 / A'$, we get an extension C of A'' by B and diagram (6) of no. 1 shows that the image of C in $\text{Ext}(A, B)$ is indeed C_1, which finishes the proof. $\square$

3. Other exact sequences

Let $0 \to B' \to B \to B'' \to 0$ be a strictly exact sequence. If A is an algebraic group and if $\varphi \in \text{Hom}(A, B'')$, then $\varphi^*(B) \in \text{Ext}(A, B')$. Putting $d(\varphi) = \varphi^*(B)$, we thus define a homomorphism

$$d : \text{Hom}(A, B'') \to \text{Ext}(A, B').$$

Proposition 3. *Let* $0 \to B' \to B \to B'' \to 0$ *be a strictly exact sequence and let* A *be an algebraic group. The sequence*

$$0 \to \text{Hom}(A, B') \to \text{Hom}(A, B) \to \text{Hom}(A, B'') \xrightarrow{d}$$
$$\text{Ext}(A, B') \to \text{Ext}(A, B) \to \text{Ext}(A, B'') \quad (12)$$

is exact.

(The homorphisms which figure in this sequence are, with the exception of d, those defined canonically by $B' \to B$ and $B \to B''$.)

The proof consists in a series of checks completely analogous to those of the preceding no. which we leave to the reader.

Remarks. 1) In characteristic $p \neq 0$, there exists an exact sequence analogous to the sequence (11), but relative to a purely inseparable isogeny of height 1, $A'' = A/\mathfrak{n}$, where $\mathfrak{n}$ denotes a vector subspace of t_A stable for the p-th power ([78], no. 8):

$$0 \to \operatorname{Hom}(A/\mathfrak{n}, B) \to \operatorname{Hom}(A, B) \to \operatorname{Hom}(\mathfrak{n}, t_B) \to$$
$$\operatorname{Ext}(A/\mathfrak{n}, B) \to \operatorname{Ext}(A, B) \to \operatorname{Ext}(\mathfrak{n}, t_B). \quad (13)$$

2) The category $\mathcal{C}_A$ of (commutative) algebraic groups is an additive category. It is an *Abelian* category (in the sense of Grothendieck [27]) when the characteristic of k is 0 and only in this case. Furthermore, $\mathcal{C}_A$ contains neither enough injectives nor enough projectives. This is what prevented us from defining $\operatorname{Ext}(A, B)$ by the general methods of homological algebra (Cartan-Eilenberg [10], Grothendieck [27]). In any case, it should be possible to adapt to $\mathcal{C}_A$ the method of Yoneda [100] and to define $\operatorname{Ext}^q(A, B)$, $q \geq 2$, permitting one to extend the exact sequences (11) and (12); see also no. 23.

4. Factor systems

Let A and B be two commutative groups. We recall that any map $f : A \times A \to B$ satisfying the identity

$$f(y, z) - f(x + y, z) + f(x, y + z) - f(x, y) = 0, \quad x,\ y,\ z \in A \quad (14)$$

is called a *factor system* on A with values in B.

If $g : A \to B$ is any map, the function δg defined by the formula

$$\delta g(x, y) = g(x + y) - g(x) - g(y) \quad (15)$$

is a factor system; such a system is called *trivial*.

The group of classes of factor systems modulo the trivial factor systems is denoted $H^2(A, B)$.

A system of factors f is called *symmetric* if it satisfies the identity

$$f(x, y) = f(y, x). \quad (16)$$

The classes of symmetric factor systems form a subgroup $H^2(A, B)_s$ of $H^2(A, B)$.

Well-known (and essentially trivial) arguments show that $H^2(A, B)$ is isomorphic to the group of classes of *central* extensions of A by B and the subgroup $H^2(A, B)_s$ corresponds to *commutative* extensions, cf. Cartan-Eilenberg [10], chap. XIV, §4. (All this is more generally valid without supposing that A is commutative.)

Now suppose that A and B are (commutative as always) algebraic groups. To simplify matters we will assume in addition that they are connected. A *rational* map $f : A \times A \to B$ satisfying the identity (14) will be called a rational factor system; such a system will be called trivial if there exists a rational map $g : A \to B$ such that $\delta g = f$. The classes of rational factor systems form a group that we denote $H^2_{rat}(A, B)$ and the symmetric factor systems form a subgroup, denoted $H^2_{rat}(A, B)_s$. Instead of considering rational maps, one could consider *regular* maps; one thus defines the groups $H^2_{reg}(A, B)$ and $H^2_{reg}(A, B)_s$.

Proposition 4.
a) *The group $H^2_{reg}(A, B)_s$ is isomorphic to the subgroup of* $\mathrm{Ext}(A, B)$ *given by extensions which admit a regular section.*
b) *The group $H^2_{rat}(A, B)_s$ is isomorphic to the subgroup of* $\mathrm{Ext}(A, B)$ *given by extensions which admit a rational section.*

PROOF OF a). If $C \in \mathrm{Ext}(A, B)$ has a regular section $s : A \to C$, put $f(x, y) = s(x + y) - s(x) - s(y)$. The map f is a regular map from A to the kernel of $C \to A$, that is to say to B. It satisfies the identities (14) and (16). Furthermore, changing the section s amounts to adding a trivial factor system to it. Denoting by $\mathrm{Ext}(A, B)_*$ the subset of $\mathrm{Ext}(A, B)$ formed by the extensions having a regular section, we have thus defined a map $\theta : \mathrm{Ext}(A, B)_* \to H^2_{reg}(A, B)_s$. One checks directly that $\mathrm{Ext}(A, B)_*$ is a subgroup of $\mathrm{Ext}(A, B)$ and that θ is a homomorphism. If $\theta(C) = 0$, the extension C has a section s which is a homorphism, in other words C is a trivial extension; thus θ is *injective*. On the other hand, let f be a symmetric factor system and define a composition law on $A \times B$ by setting

$$(x, b) * (x', b') = (x + x', b + b' + f(x, x')). \tag{17}$$

One immediately checks that this composition law makes $A \times B$ a commutative algebraic group which is an extension of A by B and which corresponds to the factor system f. Thus θ is *bijective*.

PROOF OF b). Let $\mathrm{Ext}(A, B)_{**}$ be the subset of $\mathrm{Ext}(A, B)$ formed by the extensions having a rational section. One checks that it is a subgroup, and one defines as above a homomorphism $\theta : \mathrm{Ext}(A, B)_{**} \to H^2_{rat}(A, B)_s$. The kernel of θ is formed by the extensions C which admit a rational section $s : A \to C$ which is a homomorphism. Such a section is necessarily regular (cf. for example chap. V, no, 5, lemma 6), which means that $C = 0$ and θ is *injective*. To show the surjectivity of θ, let f be a symmetric factor system and define a law of composition on $A \times B$ by the formula (17); this law makes $A \times B$ a "birational group". According to the results of Weil cited in chap. V, no. 5, there exists an algebraic group C birationally isomorphic to $A \times B$ endowed with the preceding law of composition. This group is connected and commutative (because the factor system f is symmetric). The projection $A \times B \to A$ defines a surjective homomorphism $C \to A$. If b is a generic point of B, we choose $a \in A$ and $b' \in B$ generic and independent. Denoting by $F : A \times B \to C$ the birational isomorphism introduced above, we can put $\varphi(b) = F(a, b + b') - F(a, b')$: this makes sense because $(a, b + b')$ and (a, b') are generic points of $A \times B$. Then one immediately checks that $\varphi(b)$ does not depend on the choice of (a, b'), that it is a homomorphism from B to C, and that the sequence $0 \to B \to C \to A \to 0$ is strictly exact. Furthermore, the factor system corresponding to C is equal to f, which shows that θ is *bijective*.   □

Corollary. *The canonical homomorphism* $H^2_{reg}(A, B)_s \to H^2_{rat}(A, B)_s$ *is injective.*

PROOF. Indeed, these groups are identified with subgroups of $\mathrm{Ext}(A, B)$.
                                                                                    □

Remarks. 1) If we no longer suppose that A is commutative, the preceding argument shows that $H^2_{reg}(A, B)$ (resp. $H^2_{rat}(A, B)$) is identified with the group of classes of *central* extensions of A by B which admit a regular section (resp. a rational section).

2) The above corollary can be checked directly: if $g : A \to B$ is a rational map such that $g(x + y) - g(x) - g(y) = f(x, y)$ is regular on $A \times A$, the map g is regular on a non-empty open U of A, thus also on $U + U = A$ by virtue of the preceding identity.

5. The principal fiber space defined by an extension

Let, as in the preceding no., A and B be two (commutative) connected algebraic groups. Let $C \in \mathrm{Ext}(A, B)$ be an extension admitting a rational section $s : A \to C$. This section is regular on a non-empty open U of A; since the translates of U cover A, there exists a cover $\{U_i\}$ of A and regular sections s_i of C over the U_i. The restriction of C to U_i is biregularly isomorphic to $U_i \times B$, which means that C can be considered as a *principal fiber space* with base A and structure group B. If $\mathcal{B}_A$ denotes the sheaf of germs of regular maps of A to B, the group $H^1(A, \mathcal{B}_A)$ is just the group of classes of fiber spaces of this type. The element $c \in H^1(A, \mathcal{B}_A)$ corresponding to C is defined by the 1-cocycle $b_{ij} = s_j - s_i$. Taking into account proposition 4, we have defined a canonical map

$$\pi : H^2_{rat}(A, B)_s \to H^1(A, \mathcal{B}_A).$$

Proposition 5. *The map π is a homomorphism and its kernel is equal to* $H^2_{reg}(A, B)_s$.

PROOF. The fact that π is a homomorphism can be checked by a direct computation (or follows from the fact that π is "functorial", that is to say it commutes with the homomorphisms f_* and g^* defined by $f : B \to B'$ and $g : A' \to A$). The kernel of π is formed by extensions C which are *trivial* fiber spaces, i.e., which admit a regular section. According to proposition 4, this kernel is $H^2_{reg}(A, B)_s$.   □

In §3 we will determine the image of π when A is an *Abelian variety* and B is a *linear* group; in particular, we will see that this image is equal to $H^1(A, \mathcal{B}_A)$ when B is *unipotent*.

6. The case of linear groups

As in the two preceding nos., the groups A and B are assumed to be *connected*.

Proposition 6. *If B is linear, then $H^2_{rat}(A, B)_s = \operatorname{Ext}(A, B)$.*

PROOF. In view of proposition 4, we must show that every extension C of A by B has a rational section s. Let x be a generic point of A over k and let B_x be its inverse image in C. The variety B_x is a principal homogeneous space for B, defined over the field $k(x)$. To say that C has a rational section s amounts to saying that B_x has a point $s(x)$ rational over over $k(x)$, i.e., that B_x is a *trivial* principal homogeneous space.

On the other hand, in view of the structure of commutative linear groups (Borel [9], or Rosenlicht [66]) the group B admits a composition series whose succesive quotients are isomorphic either to the multiplicative group $\mathbf{G}_m$ or to the additive group $\mathbf{G}_a$. Proposition 6 is thus a result of the following lemma (Rosenlicht [66]):

Lemma 2. *Let K be a field and let B be a K-group admitting a composition series (defined over K) whose succesive quotients are K-isomorphic to $\mathbf{G}_m$ or $\mathbf{G}_a$. Every principal homogeneous K-space over B is then trivial.*

PROOF. Let K_s be the separable closure of K and let $\mathfrak{g}_s$ be the Galois group of K_s/K operating in the natural way on the group B_s of points of B rational over K_s. One knows (Lang-Tate [54], cf. chap. V, no. 21) that the group of principal homogeneous K-spaces over B is isomorphic to $H^1(\mathfrak{g}_s, B_s)$; thus everything comes down to showing that this last group is 0. By induction on dim B, we can suppose that $B = \mathbf{G}_m$ or $B = \mathbf{G}_a$. In the first case, $B_s = K_s^*$, and the vanishing of $H^1(\mathfrak{g}_s, K_s^*)$ is nothing other than the classical "theorem 90" of Hilbert. In the second case, $B_s = K_s$ and the vanishing of $H^1(\mathfrak{g}_s, K_s)$ (and the higher cohomology groups) is no less classical; it follows, for example, from the normal basis theorem. $\square$

When A itself is linear, we can go further:

Proposition 7. *If A and B are linear, then $H^2_{reg}(A, B)_s = \operatorname{Ext}(A, B)$.*

PROOF. We know (cf. chap. III, no. 7) that every connected linear group is the product of a unipotent group with some groups $\mathbf{G}_m$. In view of the additivity of the functor Ext, we are reduced to checking the proposition when A and B are equal either to U or to $\mathbf{G}_m$. In fact,

$$\text{Ext}(\mathbf{G}_m, U) = \text{Ext}(U, \mathbf{G}_m) = \text{Ext}(\mathbf{G}_m, \mathbf{G}_m) = 0.$$

This is clear for the first two groups and, for the third, it comes from the structure of tori. Thus we are finally reduced to the case where A and B are *unipotent*. In view of propositions 5 and 6, it suffices to show that $H^1(A, \mathcal{B}_A) = 0$. We argue by induction on $n = \dim B$. If $n = 1$, we have $B = \mathbf{G}_a$, whence $\mathcal{B}_A = \mathcal{O}_A$ the sheaf of local rings of A, and $H^1(A, \mathcal{O}_A) = 0$ from the fact that A is an affine variety (cf. FAC, chap. II, §3). If $n > 1$, one knows (Rosenlicht [66]) that B has a connected subgroup B' such that $B/B' = \mathbf{G}_a$. From the fact that the extension B has a rational section (prop. 6), we have an exact sequence of sheaves

$$0 \to \mathcal{B}'_A \to \mathcal{B}_A \to \mathcal{O}_A \to 0 \qquad (18)$$

and an exact cohomology sequence

$$H^1(A, \mathcal{B}'_A) \to H^1(A, \mathcal{B}_A) \to H^1(A, \mathcal{O}_A). \qquad (19)$$

According to the induction hypothesis, the two end groups vanish, thus so does $H^1(A, \mathcal{B}_A)$.   □

Corollary. *Every connected unipotent group is biregularly isomorphic (as algebraic variety) to an affine space.*

PROOF. Let U be such a group and let us argue by induction on $\dim U$. If $\dim U = 1$, then $U = \mathbf{G}_a$. If $\dim U > 1$, there is a connected subgroup U' of U such that $U/U' = \mathbf{G}_a$. According to the preceding proposition applied to $\text{Ext}(\mathbf{G}_a, U')$, the underlying variety of U is biregularly isomorphic to $\mathbf{G}_a \times U'$, whence the result by virtue of the induction hypothesis.   □

Remarks. 1) According to our general conventions, the group U was supposed *commutative*. In fact, this hypothesis is unneccesary, as one can see by reconsidering the preceding proof.

2) The above corollary can be made a little more precise by showing that one can find in the unipotent group U coordinates $(x_1, \ldots, x_n)$ such that the sum $\vec{z} = (z_1, \ldots, z_n)$ of two elements $\vec{x} = (x_1, \ldots, x_n)$ and $\vec{y} = (y_1, \ldots, y_n)$ is given by formulas of the type:

$$z_i = x_i + y_i + P_i(x_1, y_1, \ldots, x_{i-1}, y_{i-1}) \quad (1 \le i \le n) \qquad (20)$$

where the P_i are polynomials.

§2. Structure of (commutative) connected unipotent groups

7. The group $\mathrm{Ext}(\mathbf{G}_a, \mathbf{G}_a)$

Every connected unipotent group is a multiple extension of groups of the type $\mathbf{G}_a$; in order to determine the structure of these groups the first thing to do is to make explicit the group $\mathrm{Ext}(\mathbf{G}_a, \mathbf{G}_a)$. According to proposition 7, this group is equal to $H^2_{reg}(\mathbf{G}_a, \mathbf{G}_a)_s$, the group of classes of regular symmetric factor systems. As a regular map from $\mathbf{G}_a \times \mathbf{G}_a$ to $\mathbf{G}_a$ is nothing other than a *polynomial* $f(x, y)$ *in two variables*, we are thus reduced to determining which of these polynomials are symmetric and satisfy the identity (14). In characteristic $p > 0$, here is a non-trivial example of such a polynomial:

$$F(x, y) = \frac{1}{p}(x^p + y^p - (x + y)^p). \tag{21}$$

Proposition 8. *In characteristic* 0, $H^2_{reg}(\mathbf{G}_a, \mathbf{G}_a) = 0$. *In characteristic* $p > 0$, *the k-vector space* $H^2_{reg}(\mathbf{G}_a, \mathbf{G}_a)_s$ *admits for a basis the p^i-th powers* $(i = 0, 1, \dots)$ *of the factor system F defined by the formula* (21).

PROOF. Writing $f(x, y)$ in the form $\sum a_{i,j} x^i y^j$, formula (14) translates into identities for the $a_{i,j}$ which allow one to determine explicity all factor systems (in particular those which are symmetric). For the details of the computation, see Lazard [55], §III. □

Corollary. *In characteristic* 0, *every (commutative) connected unipotent group is isomorphic to a product of groups* $\mathbf{G}_a$.

PROOF. Let U be such a group; we argue by induction on $n = \dim U$, the case $n = 0$ being trivial. If $n \geq 1$, the group U contains a connected subgroup U' such that $U/U' = \mathbf{G}_a$; in view of the induction hypothesis, $U' = (\mathbf{G}_a)^{n-1}$, thus $\mathrm{Ext}(\mathbf{G}_a, U') = 0$ since $\mathrm{Ext}(\mathbf{G}_a, \mathbf{G}_a) = 0$ by the preceding proposition. Thus $U = U' \times \mathbf{G}_a = (\mathbf{G}_a)^n$ □

From here until no. 12 *we suppose that the characteristic p of k is > 0.*

8. Witt groups

The aim of this no. is to fix a certain number of notations which will be used in the rest of the §.

We denote by W_n the *Witt group* of dimension n; an element $\vec{x} \in W_n$ is defined by n components $(x_0, \dots x_{n-1})$, $x_i \in k$; the composition law is given by formulas of the type (20); the polynomials figuring in these formulas are constructed following a procedure indicated by Witt [99].

The groups W_n are connected by the following three operations:

a) The Frobenius homomorphism $F : W_n \to W_n$ which maps $(x_0, \ldots x_{n-1})$ to $(x_0^p, \ldots x_{n-1}^p)$ (cf. chap. VI, no. 1).
b) The shift homomorphism $V : W_n \to W_{n+1}$ which maps $(x_0, \ldots x_{n-1})$ to $(0, x_0, \ldots x_{n-1})$.
c) The restriction homomorphism $R : W_{n+1} \to W_n$ which maps $(x_0, \ldots x_n)$ to $(x_0, \ldots x_{n-1})$.

These homomorphisms commute with each other. Their product is equal to multiplication by p.

If n and m are two integers ≥ 1 there is a strictly exact sequence

$$0 \to W_m \xrightarrow{V^n} W_{n+m} \xrightarrow{R^m} W_n \to 0. \tag{22}$$

The corresponding element of $\mathrm{Ext}(W_n, W_m)$ will be denoted α_n^m.

It is easy to determine the effect of the operations V and R on the α_n^m. For example, the commutative diagram

$$
\begin{array}{ccccccccc}
0 & \longrightarrow & W_m & \longrightarrow & W_{n+m} & \longrightarrow & W_n & \longrightarrow & 0 \\
& & \downarrow{\scriptstyle R} & & \downarrow{\scriptstyle R} & & \downarrow{\scriptstyle id.} & & \\
0 & \longrightarrow & W_{m-1} & \longrightarrow & W_{n+m-1} & \longrightarrow & W_n & \longrightarrow & 0
\end{array}
$$

shows that

$$R\alpha_n^m = \alpha_n^{m-1}. \tag{23}$$

One checks analogously the formulas

$$V\alpha_n^m = \alpha_{n-1}^{m+1} R \quad \text{and} \quad \alpha_n^m V = \alpha_{n-1}^m. \tag{24}$$

(In these formulas, we use the notation $V\alpha_n^m$ and $\alpha_n^m V$ to denote $V_*(\alpha_n^m)$ and $V^*(\alpha_n^m)$ respectively, cf. no. 1.)

We denote by A_n the ring of endomorphisms of the algebraic group W_n. The operations fC and Cg give $\mathrm{Ext}(W_n, W_m)$ the structure of a right module over A_n and a left module over A_m and these two structures are compatible.

When $n = 1$, the group W_1 reduces to $\mathbf{G}_a$. The exact sequences (22) show that W_n is a multiple extension of groups $\mathbf{G}_a$ and is thus a unipotent group. Identifying W_i with a subgroup of W_n by means of V^{n-i}, we have $W_i = p^{n-i}W_n$; the W_i are the only connected subgroups of W_n.

9. Lemmas

Lemma 3. *Every element* $x \in \text{Ext}(\mathbf{G}_a, \mathbf{G}_a)$ *can be written uniquely as* $x = \varphi \alpha_1^1$ *(resp.* $\alpha_1^1 \psi$*), with* φ*,* $\psi \in A_1$.

PROOF. The element $\alpha_1^1 \in \text{Ext}(\mathbf{G}_a, \mathbf{G}_a)$ corresponds to the factor system F defined by the formula (21). According to proposition 8, the element x corresponds to a factor system of the form

$$f = \sum a_i F^{p^i}, \qquad \text{with } a_i \in k.$$

On the other hand, every endomorphism φ of $\mathbf{G}_a$ can be written uniquely in the form

$$\varphi(t) = \sum b_i t^{p^i}.$$

We have $x = \varphi \alpha_1^1$ if and only if $b_i = a_i$ for all i, which proves the existence and uniqueness of φ. Similarly, writing ψ in the form

$$\psi(t) = \sum c_i t^{p^i},$$

the c_i are determined by the equation $c_i^p = a_i$. $\square$

Lemma 4. $\text{Ext}(W_n, \mathbf{G}_a)$ *is a free left* A_1*-module with basis* α_n^1.

PROOF. We argue by induction on n. For $n = 1$, $W_1 = \mathbf{G}_a$ and we apply the part of lemma 3 related to φ. For $n \geq 2$, we use the exact sequence of Ext (11) associated to the extension $W_n / \mathbf{G}_a = W_{n-1}$. Thus we get the exact sequence

$$\text{Ext}(W_{n-1}, \mathbf{G}_a) \xrightarrow{\lambda} \text{Ext}(W_n, \mathbf{G}_a) \xrightarrow{\mu} \text{Ext}(\mathbf{G}_a, \mathbf{G}_a). \qquad (25)$$

We have $\lambda(x) = xR$ and $\mu(y) = yV^{n-1}$; thus λ and μ are homomorphisms for the structure of left A_1-module. According to (24), $\lambda(\alpha_{n-1}^1) = \alpha_{n-1}^1 R = 0$, and as α_{n-1}^1 generates $\text{Ext}(W_{n-1}, \mathbf{G}_a)$ by the induction hypothesis, this implies $\lambda = 0$. Thus μ is injective. Furthermore, according to (24), $\mu(\alpha_n^1) = \alpha_n^1 V^{n-1} = \alpha_1^1$ and, according to lemma 3, α_1^1 is a basis of $\text{Ext}(\mathbf{G}_a, \mathbf{G}_a)$. It follows that μ is bijective, and that α_n^1 is indeed a basis of $\text{Ext}(W_n, \mathbf{G}_a)$ for its structure of left A_1-module. $\square$

Lemma 4'. $\text{Ext}(\mathbf{G}_a, W_n)$ *is a free right* A_1*-module with basis* α_1^n.

PROOF. The proof is identical to that of lemma 4, except that one applies the part of lemma 3 related to ψ, and one uses the exact sequence (12) associated to the extension $W_n / W_{n-1} = \mathbf{G}_a$:

$$\text{Ext}(\mathbf{G}_a, W_{n-1}) \xrightarrow{\lambda} \text{Ext}(\mathbf{G}_a, W_n) \xrightarrow{\mu} \text{Ext}(\mathbf{G}_a, \mathbf{G}_a). \qquad (26)$$

Here again, one shows that $\lambda = 0$ and that μ is an isomorphism. $\square$

Lemma 5. *Let $n \geq 0$. For every $\varphi \in A_1$, there exist Φ, $\Phi' \in A_{n+1}$ such that $\varphi R^n = R^n \Phi$ and $\Phi' V^n = V^n \varphi$.*

PROOF. By linearity, we can restrict to the case where $\varphi(t) = \lambda t^{p^i}$ with $\lambda \in k$. Choose $\vec{w} \in W_{n+1}$ such that $R^n \vec{w} = \lambda$ and define Φ by the formula

$$\Phi(\vec{x}) = \vec{w}.F^i(\vec{x}), \tag{27}$$

where the product is taken in the sense of the *ring* of Witt vectors (cf. Witt [99]). Then $R^n \Phi(\vec{x}) = \lambda.R^n F^i(\vec{x}) = \lambda.F^i R^n(\vec{x}) = \varphi R^n(\vec{x})$.

Similarly, define Φ' by the formula

$$\Phi'(\vec{x}) = \vec{w'}.F^i(\vec{x}), \qquad \text{with } R^n F^n \vec{w'} = \lambda. \tag{28}$$

$\square$

Lemma 6. *Every element x of $\mathrm{Ext}(W_n, \mathbf{G}_a)$ can be written $x = \alpha_n^1 f$, with $f \in A_n$. One has $\alpha_n^1 f = 0$ if and only if f is not an isogeny.*

PROOF. We have $xV^{n-1} \in \mathrm{Ext}(\mathbf{G}_a, \mathbf{G}_a)$, which shows (lemma 3) that we can write

$$xV^{n-1} = \alpha_1^1 \psi, \qquad \text{with } \psi \in A_1.$$

According to lemma 5, there exists $f \in A_n$ such that $fV^{n-1} = V^{n-1}\psi$. Thus

$$xV^{n-1} = \alpha_1^1 \psi = \alpha_n^1 V^{n-1} \psi = \alpha_n^1 f V^{n-1}.$$

But we saw in the proof of lemma 4 that the homomorphism μ which maps x to xV^{n-1} is bijective. The relation $xV^{n-1} = \alpha_n^1 f V^{n-1}$ thus implies $x = \alpha_n^1 f$.

Furthermore, $x = 0$ is equivalent to $xV^{n-1} = \alpha_1^1 \psi = 0$, that is to say to $\psi = 0$ (lemma 3); according to the formula $fV^{n-1} = V^{n-1}\psi$, this is equivalent to $fV^{n-1} = 0$, that is to say to $\mathrm{Ker}(f) \supset \mathbf{G}_a$, thus to $\dim \mathrm{Ker}(f) \geq 1$, as was to be shown. $\square$

(We write $\mathrm{Ker}(f)$ for the kernel of f.)

Lemma 6'. *Every element x of $\mathrm{Ext}(\mathbf{G}_a, W_n)$ can be written $x = g\alpha_1^n$, with $g \in A_n$. One has $g\alpha_1^n = 0$ if and only if g is not an isogeny.*

The proof is analogous to that of lemma 6.

Lemma 7. *If $m \geq n$, every element $x \in \mathrm{Ext}(W_n, \mathbf{G}_a)$ can be written $x = \alpha_m^n f$, with $f \in \mathrm{Hom}(W_n, W_m)$.*

PROOF. According to lemma 6, $x = \alpha_n^1 f_1$ with $f_1 \in A_n$. As $\alpha_n^1 = \alpha_m^1 V^{m-n}$, this can be written $x = \alpha_m^1 V^{m-n} f_1$, and we put $f = V^{m-n} f_1$. $\square$

Lemma 7'. *If $m \geq n$, every element $x \in \text{Ext}(\mathbf{G}_a, W_n)$ can be written $x = g\alpha_1^m$, with $g \in \text{Hom}(W_m, W_n)$.*

The proof is analogous to that of lemma 7.

10. Isogenies with a product of Witt groups

Let G be a connected unipotent group. Since G is a multiple extension of groups $\mathbf{G}_a$, there exists an integer $n \geq 0$ such that $p^n.x = 0$ for all $x \in G$. The smallest power of p satisfying this condition is called the *period* of G; if $n = \dim G$, the period of G is $\leq p^n$. Conversely:

Proposition 9. *Let G be a (commutative) connected unipotent group of dimension n. The following three conditions are equivalent:*

(i)   *The period of G is equal to p^n.*
(ii) *There exists an isogeny $f : W_n \to G$.*
(ii)' *There exists an isogeny $f' : G \to W_n$.*

PROOF. Because the period is invariant by isogeny, we have the implications (ii)⇒(i) and (ii)'⇒(i). Let us show by induction on n that (i)⇒(ii). If $n = 1$, this is trivial. If $n \geq 2$, G can be considered as an extension of a group G_1 of dimension $n - 1$ by the group $\mathbf{G}_a$. The period of G_1 is necessarily p^{n-1}, and according to the induction hypothesis there exists an isogeny $g : W_{n-1} \to G_1$. Then $g^*(G) \in \text{Ext}(W_{n-1}, \mathbf{G}_a)$; by virtue of lemma 4, there exists $\varphi \in A_1$ such that $g^*(G) = \varphi_*(W_n)$. According to no. 1, this is equivalent to saying that there exists a homomorphism $f : W_n \to G$ making a commutative diagram

$$
\begin{array}{ccccccccc}
0 & \longrightarrow & \mathbf{G}_a & \longrightarrow & W_n & \longrightarrow & W_{n-1} & \longrightarrow & 0 \\
& & \varphi\downarrow & & f\downarrow & & g\downarrow & & \\
0 & \longrightarrow & \mathbf{G}_a & \longrightarrow & G & \longrightarrow & G_1 & \longrightarrow & 0.
\end{array}
\tag{26}
$$

We have $\varphi \neq 0$, for if φ were null, the homomorphism f would define by passage to the quotient a homorphism from W_{n-1} to G and the group G would be isogenous to $\mathbf{G}_a \times W_{n-1}$, thus of period p^{n-1}. But since g is an isogeny, the same is true of f, which proves (i)⇒(ii).

The implication (i)⇒(ii)' can be checked in an analogous manner, by considering G as an extension of $\mathbf{G}_a$ by a group G_1 of dimension $n - 1$ and applying lemma 4' instead of lemma 4. □

Proposition 10. *Let $W = \prod W_{n_i}$ be a product of Witt groups, and let G be a connected unipotent group. The following conditions are equivalent:*

(a) *There exists an isogeny $f : G \to W$.*
(a)' *There exists an isogeny $g : W \to G$.*

PROOF. Suppose that (a) is true, and let G'_i be the inverse images in G of the factors W_{n_i} of W; for every i, let G_i be the connected component of the identity of G'_i. The map $f_i : G_i \to W_{n_i}$ defined by f is an isogeny. According to proposition 9, there thus exists an isogeny $g_i : W_{n_i} \to G_i$ and the map $g : W \to G$, which is the sum of the g_i, is clearly an isogeny, which proves (a)$\Rightarrow$(a)'. One argues similarly to check (a)'$\Rightarrow$(a).   □

We say that G is *isogenous to W* if the equivalent conditions (a) and (a)' are satisfied. If $G_1 \to G_2$ is an isogeny, it is clear that G_1 is isogenous to W if and only if G_2 is.

Theorem 1. *Every (commutative) connected unipotent group is isogenous to a product of Witt groups.*

PROOF. Let G be a connected unipotent group of dimension r. We argue by induction on r, the case $r = 1$ being trivial. The group G is an extension of a group G_1 of dimension $r-1$ by the group $\mathbf{G}_a$. In view of the induction hypothesis, there exists an isogeny

$$f : \prod_{i=1}^{i=k} W_{n_i} \to G_1.$$

Put $W = \prod_{i=1}^{i=k} W_{n_i}$. The group $f^*(G)$ is an extension of W by $\mathbf{G}_a$, and is isogenous to G; it thus suffices to show that this group is isogenous to a product of Witt groups, in other words we are reduced to the case where $W = G_1$.

In this case, the extension G is defined by a family of elements $\gamma_i \in \text{Ext}(W_{n_i}, \mathbf{G}_a)$. Suppose $n_1 \geq n_i$ for all i and let W' be the product of the W_{n_i} for $i \geq 2$. We are going to distinguish two cases:

i) $\gamma_1 = 0$. The group G is then the product of W_{n_1} and of the extension of W' by $\mathbf{G}_a$ defined by the system (γ_i), $i \geq 2$. The induction hypothesis shows that G is isogenous to a product of Witt groups.

ii) $\gamma_1 \neq 0$. Let $\beta = (\beta_i) \in \text{Ext}(W, \mathbf{G}_a)$ be the element defined by $\beta_1 = \alpha^1_{n_1}$, $\beta_i = 0$ if $i \geq 2$. The extension G' corresponding to β is nothing other than the product $W_{n_1+1} \times W'$. We are going to show the existence of an isogeny $\varphi : W \to W$ such that $\varphi^*(G')$ is isomorphic to G; it will follow from this that G is isogenous to G', which is a product of Witt groups.

According to lemma 7, there exists $f_i \in \text{Hom}(W_{n_i}, W_{n_1})$ such that $\gamma_i = \alpha^1_{n_1} f_i$. Thus we define $\varphi : W \to W$ by the formula

$$\varphi(w_1, w_2, \ldots, w_k) = (f_1(w_1) + \cdots + f_k(w_k), w_2, \ldots, w_k).$$

An immediate computation gives $\varphi^*(\beta) = \gamma$, denoting by γ the element of $\text{Ext}(W, \mathbf{G}_a)$ defined by G. Further, to show that φ is surjective it suffices to see that f_1 is, which follows from lemma 6 since $\alpha^1_{n_1} f_1 \neq 0$.   □

Remarks. 1) In an isogeny between G and $\prod W_{n_i}$, the integers n_i are *unique* (up to order). Indeed, denoting by u_n the number of those that are equal to n, one immediately checks the formula

$$u_n = \dim(p^{n-1}G/p^n G) - \dim(p^n G/p^{n+1}G). \qquad (27)$$

2) Theorem 1 is analogous to the structure theorem for Abelian p-groups. This analogy can be pursued further; for example, one can show that a subgroup H of a connected unipotent group G is a "quasi-direct factor" in G (direct factor up to an isogeny) if and only if this subgroup is "quasi-pure" in other words if $\dim p^n H = \dim(H \cap p^n G)$ for all n.

11. Structure of connected unipotent groups: particular cases

Theorem 1 gives the structure of connected unipotent groups only up to an isogeny. In certain cases one can go further. For example:

Proposition 11. *Every (commutative) connected unipotent group of period p is isomorphic to a product of groups $\mathbf{G}_a$.*

PROOF. We argue by induction on $r = \dim G$, the case $r = 1$ being trivial. The group G is an extension of a group G_1 by the group $\mathbf{G}_a$. In view of the induction hypothesis, $G_1 = (\mathbf{G}_a)^{r-1}$, and G is defined by $r - 1$ elements $\gamma_i \in \mathrm{Ext}(\mathbf{G}_a, \mathbf{G}_a)$. If one of the γ_i were non-zero, for example γ_1, the proof of theorem 1 would show that G is isogenous to $W_2 \times (\mathbf{G}_a)^{r-2}$, which is absurd because G has period p. All the γ_i being zero, we have $G = \mathbf{G}_a \times G_1 = (\mathbf{G}_a)^r$, as was to be shown. (For a direct proof, see Rosenlicht [68], prop. 2.) □

When G is of dimension 2, we can easily give a complete system of invariants. Eliminating the case where $G = (\mathbf{G}_a)^2$, put $G' = pG$ and denote by G'' the subgroup of G formed by elements $x \in G$ such that $px = 0$. The group G'' is of dimension 1 and its connected component of the identity is G'. The quotient G''/G' is a finite subgroup of G/G', a group which one can identify with $\mathbf{G}_a$. Thus we get a *first invariant*, which is a finite subgroup of $\mathbf{G}_a$, determined up to a non-zero scalar multiplication. On the other hand, the map $x \to px$ defines by passage to the quotient a bijective homomorphism $G/G'' \to G'$ whose tangent mapping is zero; it is thus a purely inseparable isogeny, of degree p^h, with $h \geq 1$. The integer h is the *second invariant* of G.

Using lemma 3, it is not difficult to check that the invariants N and h *characterize* G up to an isomorphism, and that they can be chosen arbitrarily. The group G is *separably* (resp. *purely inseparably*) isogenous to W_2 if and only if $h = 1$ (resp. $N = 0$).

12. Other results

Theorem 2. *Every (commutative) connected unipotent group is isomorphic to a subgroup of a product of Witt groups.*

PROOF. Let G be such a group and let r be its dimension. We argue by induction on r, the case $r = 1$ being trivial. The group G is an extension of $\mathbf{G}_a$ by a group G_1 of dimension $r - 1$. According to the induction hypothesis, G_1 can be embedded in a product W of Witt groups, and G can be embedded in the corresponding extension of $\mathbf{G}_a$ by W. Thus we are reduced to the case where $G_1 = W$.

Let $W = \prod_{i=1}^{i=m} W_{n_i}$ be a decomposition of W into a product of Witt groups. The element $G \in \mathrm{Ext}(\mathbf{G}_a, W)$ is defined by a family of elements $\gamma_i \in \mathrm{Ext}(\mathbf{G}_a, W_{n_i})$. According to lemma 4', there exist $\varphi_i \in A_1$ such that $\gamma_i = \alpha_1^{n_i} \cdot \varphi_i$. Then put

$$L = \prod_{i=1}^{i=m} W_{n_i+1} \times \mathbf{G}_a.$$

We can naturally consider L as an extension of $(\mathbf{G}_a)^m \times \mathbf{G}_a$ by W; let $\beta \in \mathrm{Ext}((\mathbf{G}_a)^m \times \mathbf{G}_a, W)$ be the corresponding element. Then we define a homomorphism $\theta : \mathbf{G}_a \to (\mathbf{G}_a)^m \times \mathbf{G}_a$ by the formula

$$\theta(x) = (\varphi_1(x), \ldots, \varphi_m(x), x).$$

One immediately checks that $\beta\theta = G$. Thus there exists a homomorphism $\psi : G \to L$ making a commutative diagram

$$
\begin{array}{ccccccc}
0 & \longrightarrow & W & \longrightarrow & G & \longrightarrow & \mathbf{G}_a & \longrightarrow & 0 \\
& & \downarrow{\scriptstyle id.} & & \downarrow{\scriptstyle \psi} & & \downarrow{\scriptstyle \theta} & & \\
0 & \longrightarrow & W & \longrightarrow & L & \longrightarrow & (\mathbf{G}_a)^m \times \mathbf{G}_a & \longrightarrow & 0.
\end{array}
$$

Furthermore, θ is an embedding of $\mathbf{G}_a$ in $(\mathbf{G}_a)^m \times \mathbf{G}_a$; the same is thus also true of ψ which finishes the proof.   □

Theorem 3. *Every (commutative) connected unipotent group is isomorphic to a quotient of a product of Witt groups by a connected subgroup.*

PROOF. The proof is "dual" to the preceding one. We argue by induction on the dimension of the given group G, by considering G as an extension of a group G_1 by $\mathbf{G}_a$. Applying the induction hypothesis to G_1, we can suppose that G_1 is a product $\prod_{i=1}^{i=m} W_{n_i}$ of Witt groups. Using lemma 4, one shows, as above, that G is isomorphic to the quotient of $\prod_{i=1}^{i=m} W_{n_i+1} \times \mathbf{G}_a$ by a connected subgroup (in fact isomorphic to $(\mathbf{G}_a)^m$).   □

13. Comparison with generalized Jacobians

Let $J_\mathfrak{m}$ be the generalized Jacobian of a curve X with respect to a modulus $\mathfrak{m} = \sum n_P P$ with support S. We saw in chap. V that $J_\mathfrak{m}$ is an extension of the usual Jacobian of X by a linear group $L_\mathfrak{m}$, itself isomorphic to the product of a torus by a unipotent group $V = \prod_{P \in S} V_{(n_P)}$. We also saw, by means of the Artin-Hasse exponential, that the group V is *isomorphic* (and not just isogenous) to a product of Witt groups. We can use this fact to give a new proof of theorem 3. First we establish the following result:

Theorem 4. *Every (commutative) connected group is isomorphic to the quotient of a product of generalized Jacobians by a connected subgroup.*

PROOF. Let G be such a group, and let e be its identity element. Since e is a simple point on G, there exist curves X_i' on G which admit e as a simple point and whose tangents t_i at this point generate the tangent space t_G. If X_i denotes the normalization of X_i', the rational map $X_i \to X_i' \to G$ factors as $X_i \to J_i \to G$, where J_i is a certain generalized Jacobian of X_i. Put $J = \prod J_i$; the sum of the homomorphisms $J_i \to G$ defines a homomorphism $\theta : J \to G$. The image of t_J by θ contains the tangents t_i to the X_i'; thus $\theta(t_J) = t_G$ which shows at the same time that θ is *surjective* and that it is *separable*. Thus the group G is identified with the quotient J/N, where N denotes the kernel of θ. It remains to arrange for N to be connected, and for this we must introduce yet another generalized Jacobian:

Let N_0 be the connected component of the identity element of N, and let $G_0 = J/N_0$. We have $(J/N_0)/(N/N_0) = J/N = G$; the canonical projection $G_0 \to G$ is thus a separable isogeny. If $\dim G \geq 2$, the Bertini theorem (chap. VI, lemma 11) shows that the inverse image in G_0 of a suitable hyperplane section of G is irreducible. By induction, we deduce the existence of a curve X_1' on G whose inverse image in G_0 is irreducible (this is valid only if $\dim G \geq 1$—the case where G is of dimension 1 is at any rate trivial). If X_1 denotes the normalization of X_1', the map $X_1 \to G$ defines as above a homomorphism

$$\theta_1 : J_1 \to G$$

where J_1 is a generalized Jacobian of X_1. Let J_0 be the pull-back of G_0 by θ_1, that is to say the subgroup of $J_1 \times G_0$ formed by the pairs having the same image in G. The group J_0 is connected, for otherwise the inverse image of X_1' in G_0 would not be irreducible. Then define a homomorphism $\varphi : J \times J_1 \to G$ by putting $\varphi(j, j_1) = \theta(j) - \theta_1(j_1)$. Since $t_J \to t_G$ is surjective, the same is true of $t_{J \times J_1} \to t_G$, which shows that φ is *surjective* and *separable*. Furthermore, one checks immediately that the kernel M of φ is an extension of the group J_0 by N_0; as both are connected, M is *connected*, as was to be shown. $\square$

Corollary. *Every Abelian variety is isomorphic to the quotient of a product of (usual) Jacobians by a connected subgroup.*

PROOF. Let G be an Abelian variety; according to the preceding, $G = (\prod J_{m_i})/N$, where the J_{m_i} are generalized Jacobians and where N is connected. If L_{m_i} denotes the linear part of J_{m_i}, the homomorphism $L_{m_i} \to G$ is necessarily null, which means that N contains all the L_{m_i}. Then putting $J_i = J_{m_i}/L_{m_i}$, G is identified with a quotient $(\prod J_i)/N'$, where N' is the image of N (thus connected); as the J_i are usual Jacobians, the corollary is proved.   □

We return now to the case where G is *unipotent*. According to the corollary to prop. 7, the underlying variety of G is isomorphic to an affine space k^n. In the proof of thm. 4 one can then choose *lines* for the curves X_i' and X_1'. The corresponding Jacobians are reduced to their unipotent part, which is isomorphic to a product of Witt groups, as we recalled above. Thus we indeed get theorem 3. Moreover, we remark that, starting from this theorem, it is easy to recover theorem 1.

§3. Extensions of Abelian varieties

14. Primitive cohomology classes

Let A be an Abelian variety and let B be a connected linear group. According to prop. 6, $\mathrm{Ext}(A, B) = H^2_{rat}(A, B)_s$ and, according to no. 5, there is a homomorphism

$$\pi : H^2_{rat}(A, B)_s \to H^1(A, \mathcal{B}_A)$$

where $\mathcal{B}_A$ denotes the sheaf of germs of regular maps from A to B. We will again denote by π the homomorphism from $\mathrm{Ext}(A, B)$ to $H^1(A, \mathcal{B}_A)$ thus defined. We propose to determine the kernel and the image of π. For this we are going to need some preliminary definitions of a homological nature:

Let q be an integer ≥ 1 and, for every algebraic variety X, put $T(X) = H^q(X, \mathcal{B}_X)$. One thus defines a contravariant functor in X, which is zero when X is reduced to a point. If $f : X \to Y$ is a regular map, we denote by $f^* : T(Y) \to T(X)$ the homomorphism associated to f.

Now let X_1 and X_2 be two varieties, each endowed with a "marked" point, denoted e. Let $p_i : X_1 \times X_2 \to X_i$ $(i = 1, 2)$ be the projection to the i-th factor, and let $m_i : X_i \to X_1 \times X_2$ be the injection defined by e. Homomorphisms p_i^* and m_i^* correspond to these maps. Since $p_i \circ m_i$ is the identity and $p_i \circ m_j$, $i \neq j$ is a constant map, we have formulas

$$m_i^* \circ p_i^* = 1, \qquad m_j^* \circ p_i^* = 0 \quad \text{if } i \neq j. \tag{28}$$

These formulas mean that the homomorphism $p^* : T(X_1) \times T(X_2) \to T(X_1 \times X_2)$ defined by (p_1^*, p_2^*) admits as left inverse the homomorphism defined by the m_i^*. Thus we can *identify* $T(X_1) \times T(X_2)$ *with a direct factor of* $T(X_1 \times X_2)$. An element of the group $T(X_1 \times X_2)$ which belongs to this direct factor subgroup is called *decomposable*.

This applies in particular when $X_1 = X_2 = A$, where A is a (commutative) algebraic group. The group $T(A \times A)$ thus contains $T(A) \times T(A)$ as a direct factor. Let $x \in T(A)$ and denote by $s_A : A \times A \to A$ the composition law of A; we have $s_A^*(x) \in T(A \times A)$. Because $s_A \circ m_i$ is the identity $(i = 1, 2)$, we see that the component of $s_A^*(x)$ in the factor $T(A) \times T(A)$ is equal to (x, x). One says that x is *primitive* if

$$s_A^*(x) = (x, x) \qquad (\text{i.e., } p_1^*(x) + p_2^*(x)), \tag{29}$$

which amounts to saying that $s_A^*(x)$ is *decomposable*.

Denote by $PT(A)$ the subgroup of $T(A)$ formed by the primitive elements of $T(A)$.

Lemma 8. $PT(A)$ *is an additive functor of A.*

PROOF. First we must show that, if $\varphi : A \to C$ is a homomorphism, φ^* maps $PT(C)$ into $PT(A)$, which follows immediately from a diagram. Next we must show that $PT(A)$ is *additive* in A, that is to say that, if $\varphi, \psi, \theta \in \mathrm{Hom}(A, C)$ are such that $\theta = \varphi + \psi$, then $\theta^*(x) = \varphi^*(x) + \psi^*(x)$ for all $x \in PT(C)$. But the homomorphism θ can be factored as

$$A \xrightarrow{\chi} C \times C \xrightarrow{s_C} C$$

the homomorphism χ being that defined by the pair (φ, ψ). Then we have

$$\begin{aligned}
\theta^*(x) &= \chi^*(s_C^*(x)) \\
&= \chi^*(p_1^*(x) + p_2^*(x)) \\
&= \varphi^*(x) + \psi^*(x),
\end{aligned}$$

since $p_1 \circ \chi = \varphi$ and $p_2 \circ \chi = \psi$. □

Moreover, formula (29) shows that $PT(A)$ is the "largest" subgroup of $T(A)$ for which lemma 8 is true.

15. Comparison between $\mathrm{Ext}(A, B)$ and $H^1(A, \mathcal{B}_A)$

Theorem 5. *Let A be an Abelian variety and let B be a (commutative) connected linear group. The canonical homomorphism*

$$\pi : \mathrm{Ext}(A, B) \to H^1(A, \mathcal{B}_A) \qquad (cf.\ no.\ 14)$$

is injective and its image is the set of primitive elements of $H^1(A, \mathcal{B}_A)$.

PROOF. Let C be an extension of A by B belonging to the kernel of π. As a fiber space, C is trivial, that is to say it has a regular section $s : A \to C$. After making a translation by an element of B, we can suppose that $s(e) = e$ (denoting by e the identity element of the group considered). Since A is complete, $s(A)$ is a complete subvariety of C. Let A' be the subgroup of C generated by $s(A)$; it is also a complete group. The group $A' \cap B$ being at the same time complete and linear is necessarily finite. Because it is the kernel of the projection $A' \to A$, dim A' = dim A = dim $s(A)$ so $A' = s(A)$, since A' is irreducible and contains $s(A)$. This means that $s(A)$ is a subgroup of C, that is to say that s is a homomorphism and the extension C is trivial. We have thus checked that π is *injective*.

Now we take $C \in \mathrm{Ext}(A, B)$ and check that $x = \pi(C)$ is a primitive element of $H^1(A, \mathcal{B}_A)$. Observe first that π is functorial, that is to say that it commutes with the homomorphisms φ^* defined by $\varphi \in \mathrm{Hom}(A, A')$. Thus

$$s_A^*(x) = s_A^* \pi(C) = \pi s_A^*(C) = \pi(p_1^*(C) + p_2^*(C)) \quad \text{(cf. prop. 1)}$$
$$= p_1^* \pi(C) + p_2^* \pi(C) = p_1^*(x) + p_2^*(x)$$

which indeed shows that x is primitive.

Conversely, let x be a primitive element of $H^1(A, \mathcal{B}_A)$ and let C be a principal fiber space with base A and structural group B corresponding to x. We must show that there exists a structure of algebraic group on C which makes it an extension of A by B. Let C' be the pull-back by $s_A : A \times A \to A$. The hypothesis that x is primitive means that C' is isomorphic to the fiber space deduced from $C \times C$ by the homomorphism $s_B : B \times B \to B$. Composing the maps $C \times C \to C'$ and $C' \to C$, we get a regular map $g : C \times C \to C$ which makes a commutative diagram

$$\begin{CD}
C \times C @>g>> C \\
@VVV @VVV \\
A \times A @>s_A>> A
\end{CD}$$

(30)

and one can check the identity

$$g(c + b, c' + b') = g(c, c') + b + b' \quad (c,\ c' \in C,\ \ b,\ b' \in B). \quad (31)$$

Choose a point $e \in C$ projecting to the identity element of A. After effecting a translation by an element of B, we can suppose that $g(e, e) = e$. Everything comes down to checking that g gives C the structure of commutative algebraic group with identity element e. Formula (31) and diagram (30) will indeed show that C is then an extension of A by B. We must check:

a) *That* $g(c, e) = g(e, c) = c$ *for all* $c \in C$.

According to (30), $g(c, e) = c + h(c)$ where h is a regular map from C to B; furthermore, formula (31) shows that $h(c + b) = h(c)$ for all $b \in B$, in other words that h factors as $C \rightarrow A \rightarrow B$. As A is complete and B is linear, the map $\overline{h} : A \rightarrow C$ thus defined is constant. Furthermore, the formula $g(e,e) = e$ shows that $\overline{h}(e) = e$, whence $\overline{h}(c) = e$ for all $c \in C$, which proves the formula $g(c,e) = c$. One argues similarly for the formula $g(e,c) = c$.

b) *That $g(c,c') = g(c',c)$ for all $c, c' \in C$.*

According to (30), $g(c,c') = g(c',c) + k(c,c')$ where k is a regular map from $C \times C$ to B. Formula (31) shows that k factors as $C \times C \rightarrow A \times A \rightarrow B$ and the map $A \times A \rightarrow B$ is necessarily constant and equal to e, whence the result.

c) *That $g(c, g(c', c'')) = g(g(c,c'), c'')$ for all $c, c', c'' \in C$.*

Same argument as in a) and b).

d) *That there exists a regular map $i : C \rightarrow C$ such that $g(c, i(c)) = e$ for all $c \in C$.*

Denote by i_A (resp. i_B) the map $a \rightarrow -a$ (resp. $b \rightarrow -b$) from A (resp. B) to itself. It is clear that $i_{B*}(x) = -x$. The analogous formula $i_A^*(x) = -x$ is true because x is *primitive* (lemma 8). Since $i_A^*(x) = i_{B*}(x)$ there exists a regular map $i : C \rightarrow C$ making a commutative diagram

$$
\begin{array}{ccc}
C & \xrightarrow{\ i\ } & C \\
\downarrow & & \downarrow \\
A & \xrightarrow{\ i_A\ } & A
\end{array}
\qquad (32)
$$

and such that

$$i(c + b) = i(c) - b \quad \text{for} \quad c \in C, \ b \in B. \qquad (33)$$

We can also suppose that $i(e) = e$. One then shows, by the same argument as in a), b), and c), that $g(c, i(c)) = e$ for all $c \in C$, which finishes the proof. $\qquad \square$

16. The case $B = \mathbf{G}_m$

When B is the multiplicative group $\mathbf{G}_m$, the sheaf $\mathcal{B}_A$ is just the sheaf $\mathcal{O}^*$ of invertible elements of the sheaf of rings $\mathcal{O}_A$ of the Abelian variety A. The group $H^1(A, \mathcal{O}_A^*)$ is the group $D(A)$ of *divisor classes* of A (for linear equivalence). To say that the class of a divisor X is a primitive element of $D(A)$ means that $s_A^{-1}(X)$ is linearly equivalent on $A \times A$ to a "decomposed" divisor

$$X \times A + A \times X = p_1^{-1}(X) + p_2^{-1}(X).$$

We will write this as $X \equiv 0$ (cf. Lang, [**52**], p. 90, thm. 2). If we denote by $P(A)$ the subgroup of $D(A)$ formed by the classes of such divisors, theorem 5 gives us:

Theorem 6. *If A is an Abelian variety, the group* $\mathrm{Ext}(A, \mathbf{G}_m)$ *is canonically isomorphic to the group $P(A)$ of divisor classes such that $X \equiv 0$.*

One knows (Barsotti [5], see also [78]) that $X \equiv 0$ if and only if X is *algebraically equivalent to 0*. The group $P(A)$ is thus the Abelian group underlying the *dual variety* of A.

Remark. According to proposition 6, $\mathrm{Ext}(A, \mathbf{G}_m) = H^2_{rat}(A, \mathbf{G}_m)_s$. Thus every divisor X such that $X \equiv 0$ corresponds to a *class of rational factor systems* on A with values in $\mathbf{G}_m$. We can make explicit a system of factors belonging to this class in the following manner:

Since $s_A^{-1}(X) \sim p_1^{-1}(X) + p_2^{-1}(X)$ on $A \times A$, there exists a rational function $f(x, y)$ on $A \times A$ such that

$$(f) = s_A^{-1}(X) - p_1^{-1}(X) - p_2^{-1}(X). \tag{34}$$

The function f is the desired factor system.

17. The case $B = \mathbf{G}_a$

When B is the additive group $\mathbf{G}_a$, the sheaf $\mathcal{B}_A$ is just the sheaf $\mathcal{O}_A$ of local rings of A, and we are led to study the group $H^1(A, \mathcal{O}_A)$.

More generally, if X is any algebraic variety, we will write $H^q(X)$ in place of $H^q(X, \mathcal{O}_X)$ and we put

$$H^*(X) = \sum_{q=0}^{\infty} H^q(X).$$

The cup-product operation endows $H^*(X)$ with the structure of a graded algebra. As the multiplication in $\mathcal{O}_X$ is associative and commutative, that of $H^*(X)$ is associative and anti-commutative; it has a unit element 1 of degree 0 (for all these properties of the cup-product, we refer to the work of Godement [25], chap. II, §6).

If X and Y are two varieties, the projections p_1 and p_2 from $X \times Y$ to X and to Y define homomorphisms

$$p_1^* : H^*(X) \rightarrow H^*(X \times Y) \quad \text{and} \quad p_2^* : H^*(Y) \rightarrow H^*(X \times Y)$$

cf. no. 14. By means of the cup-product, we deduce a homomorphism

$$p_1^* \otimes p_2^* : H^*(X) \otimes H^*(Y) \rightarrow H^*(X \times Y)$$

the tensor product being taken over the base field k. One has a "Künneth formula":

Proposition 12. *The homomorphism $p_1^* \otimes p_2^*$ defined above is an isomorphism from $H^*(X) \otimes H^*(Y)$ to $H^*(X \times Y)$.*

PROOF. The proof is as in the classical case. We choose finite covers $\mathfrak{U}$ (resp. $\mathfrak{V}$) of X (resp. of Y) by open affines U_i (resp. V_i); the $W_{ij} = U_i \times V_j$ then form a finite cover $\mathfrak{W}$ of $X \times Y$ by open affines. By virtue of FAC, p. 239, thm. 4, the cohomology groups of X are those of the complex $C(\mathfrak{U}, \mathcal{O}_X)$ which is a "simplicial cochain complex" in the terminology of Godement [25], chap. I, §3.1; an analogous result holds for Y and for $X \times Y$. Furthermore,

$$\Gamma(W_{i_0 j_0, \ldots, i_p j_p}, \mathcal{O}_{X \times Y}) = \Gamma(U_{i_0, \ldots, i_p}, \mathcal{O}_X) \otimes \Gamma(V_{j_0, \ldots, j_p}, \mathcal{O}_Y)$$

since the coordinate ring of a product of affine varieties is the tensor product of the coordinate ring of these varieties. This formula means that the complex $C(\mathfrak{W}, \mathcal{O}_{X \times Y})$ is identified with the *Cartesian product* $C(\mathfrak{U}, \mathcal{O}_X) \times C(\mathfrak{V}, \mathcal{O}_Y)$ of the complexes $C(\mathfrak{U}, \mathcal{O}_X)$ and $C(\mathfrak{V}, \mathcal{O}_Y)$ (for the definition of the Cartesian product of two simplicial cochain complexes, see Godement [25], chap. I, §3.6). According to the Eilenberg-Zilber theorem (Godement, *loc. cit.*, theorem 3.10.1) the complex $C(\mathfrak{W}, \mathcal{O}_{X \times Y})$ is thus homotopically equivalent to the tensor product

$$C(\mathfrak{U}, \mathcal{O}_X) \otimes C(\mathfrak{V}, \mathcal{O}_Y).$$

Applying the (usual) Künneth formula to this last complex, we deduce that $H^*(X \times Y)$ is identified with the tensor product of $H^*(X)$ and $H^*(Y)$. Finally, the fact that this identification is given by the cup-product of p_1^* and p_2^* can be checked either by using the explicit formula giving the cup-product and comparing with the proof of the Eilenberg-Zilber theorem, or by using the definition of the cup-product by means of the diagonal map (Godement, *loc. cit.*, chap. II, §6).   □

Corollary. *If X and Y are complete and connected, $H^1(X \times Y)$ is identified with the direct sum of $H^1(X)$ and $H^1(Y)$.*

PROOF. Indeed, the preceding proposition shows that $H^1(X \times Y)$ is the direct sum of $H^1(X) \otimes H^0(Y)$ and $H^0(X) \otimes H^1(Y)$, and the hypotheses made on X and Y imply that $H^0(X) = H^0(Y) = k$.   □

Remark. The Kunneth formula holds for arbitrary coherent sheaves and not just for sheaves of local rings. Precisely, if $\mathcal{F}$ and $\mathcal{G}$ are two coherent sheaves on X and Y respectively, we define a coherent sheaf $\mathcal{F} \otimes \mathcal{G}$ on $X \times Y$ by putting

$$\mathcal{F} \otimes \mathcal{G} = \mathcal{F} \otimes \mathcal{O}_{X \times Y} \otimes \mathcal{G}$$

the two tensor products on the right hand side being taken over $\mathcal{O}_X$ and $\mathcal{O}_Y$ respectively. The proof of proposition 12 then shows that $H^*(X \times Y, \mathcal{F} \otimes \mathcal{G})$ is identified with $H^*(X, \mathcal{F}) \otimes H^*(Y, \mathcal{G})$.

Theorem 7. *If A is an Abelian variety, the group $\mathrm{Ext}(A, \mathbf{G}_a)$ is canonically isomorphic to $H^1(A, \mathcal{O}_A)$.*

PROOF. Indeed, according to the corollary to proposition 12 every element of $H^1(A \times A, \mathcal{O}_{A \times A})$ is decomposable, thus every element of $H^1(A, \mathcal{O}_A)$ is primitive, and we apply theorem 5. $\square$

Remarks. 1) We will see in no. 21 that the dimension of the k-vector space $H^1(A, \mathcal{O}_A)$ is equal to dim A.

2) If p is the characteristic of the ground field, one can use theorem 7 to determine $\mathrm{Ext}(A, \mathbf{Z}/p\mathbf{Z})$. Indeed, denoting by $\wp : \mathbf{G}_a \to \mathbf{G}_a$ the isogeny defined by the formula $\wp(\lambda) = \lambda^p - \lambda$ (cf. chap. VI), there is a strictly exact sequence

$$0 \to \mathbf{Z}/p\mathbf{Z} \to \mathbf{G}_a \xrightarrow{\wp} \mathbf{G}_a \to 0.$$

Applying the second exact sequence of Ext (§1, prop. 3), we see that $\mathrm{Ext}(A, \mathbf{Z}/p\mathbf{Z})$ is identified with the kernel of $\wp$ acting on $H^1(A, \mathcal{O}_A)$. But Artin-Schreier theory shows that the elements of this kernel correspond to unramified cyclic coverings of degree p of A (cf. [77], no. 16); thus we recover a particular case of the theorem that *every unramified covering of an Abelian variety is an isogeny* ([53], theorem 2). In the cyclic of order prime-to-p case one can make an analogous argument using the group $\mathbf{G}_m$ in place of the group $\mathbf{G}_a$ and theorem 6 in place of theorem 7; cf. Weil [89], §XI.

18. Case where B is unipotent

First we are going to generalize the corollary to proposition 12:

Proposition 13. *Let B be a (commutative) connected unipotent group. If X and Y are complete connected varieties, $H^1(X \times Y, \mathcal{B}_{X \times Y})$ is identified with the direct sum of $H^1(X, \mathcal{B}_X)$ and $H^1(Y, \mathcal{B}_Y)$.*

PROOF. We argue by induction on $n = \dim B$, the case $n = 1$ being the corollary to proposition 12. If $n > 1$, choose a connected subgroup B' of B such that $B/B' = \mathbf{G}_a$. The exact sequence of sheaves

$$0 \to \mathcal{B}'_X \to \mathcal{B}_X \to \mathcal{O}_X \to 0$$

gives rise to a cohomology exact sequence

$$H^0(X, \mathcal{B}_X) \to H^0(X, \mathcal{O}_X) \to H^1(X, \mathcal{B}'_X) \to H^1(X, \mathcal{B}_X) \to H^1(X, \mathcal{O}_X).$$

But since X is complete and connected, every regular map from X to one of the groups B', B, or $\mathbf{G}_a$ is constant. Thus $H^0(X, \mathcal{B}_X) = B$ and $H^0(X, \mathcal{O}_X) = \mathbf{G}_a$; the preceding exact sequence reduces to

$$0 \to H^1(X, \mathcal{B}'_X) \to H^1(X, \mathcal{B}_X) \to H^1(X, \mathcal{O}_X). \tag{35}$$

There are analogous results for Y and $X \times Y$. On the other hand, choosing "base points" in X and Y, the corresponding injections of X and Y into $X \times Y$ define a homomorphism

$$m^* : H^1(X \times Y, \mathcal{B}_{X \times Y}) \to H^1(X, \mathcal{B}_X) \times H^1(Y, \mathcal{B}_Y).$$

We know that m^* is surjective and has a right inverse p^* defined by the projections $X \times Y \to X$ and $X \times Y \to Y$, cf. no. 14. Everything comes down to showing that m^* is injective. But according to (35) there is a commutative diagram

$$
\begin{array}{ccccc}
0 \to & H^1(X \times Y, \mathcal{B}'_{X \times Y}) & \to & H^1(X \times Y, \mathcal{B}_{X \times Y}) & \to & H^1(X \times Y, \mathcal{O}_{X \times Y}) \\
& m^* \downarrow & & m^* \downarrow & & m^* \downarrow \\
0 \to & H^1(X, \mathcal{B}'_X) \times H^1(Y, \mathcal{B}'_Y) \to & H^1(X, \mathcal{B}_X) \times H^1(Y, \mathcal{B}_Y) & \to & H^1(X, \mathcal{O}_X) \times H^1(Y, \mathcal{O}_Y).
\end{array}
$$

In view of the induction hypothesis, the two vertical arrows on the ends are bijections; thus the middle vertical arrow is an injection, which finishes the proof.   □

Theorem 8. *Let A be an Abelian variety and let B be a (commutative) connected unipotent group. The canonical homomorphism*

$$\pi : \operatorname{Ext}(A, B) \to H^1(A, \mathcal{B}_A)$$

defined in no. 14 is bijective.

PROOF. Indeed, proposition 13 says that every element of $H^1(A \times A, \mathcal{B}_{A \times A})$ is decomposable, thus that every element of $H^1(A, \mathcal{B}_A)$ is primitive and we apply theorem 5.

§4. Cohomology of Abelian varieties

19. Cohomology of Jacobians

Let X be an irreducible, projective curve without singularities and let $\varphi : X \to J$ be the canonical map of X into its Jacobian. This map defines a homomorphism $\varphi^* : H^1(J, \mathcal{O}_J) \to H^1(X, \mathcal{O}_X)$; although φ is only defined up to a translation, φ^* is uniquely determined since translations act trivially on $H^*(J)$ by the Künneth formula.

Theorem 9. *The homomorphism $\varphi^* : H^1(J, \mathcal{O}_J) \to H^1(X, \mathcal{O}_X)$ is bijective.*

PROOF. We first remark that $H^1(J, \mathcal{O}_J) = \operatorname{Ext}(J, \mathbf{G}_a)$ according to theorem 7. As the map φ is maximal (in the sense of chap. VI, no. 13), the fact that φ^* is *injective* will follow from this more general proposition:

Proposition 14. *Let X be a complete variety, A an Abelian variety, and B a (commutative) connected linear group. If $\varphi : X \to A$ is an everywhere regular maximal map, the composed homomorphism*

$$\text{Ext}(A, B) \xrightarrow{\pi} H^1(A, \mathcal{B}_A) \xrightarrow{\varphi^*} H^1(X, \mathcal{B}_X)$$

is injective.

PROOF. Let C be an extension of A by B. To say that C belongs to the kernel of the homomorphism $\text{Ext}(A, B) \to H^1(X, \mathcal{B}_X)$ means that the pull-back fiber space of C by φ is trivial, in other words that φ factors as $X \xrightarrow{\psi} C \to A$, where ψ is a regular map. After effecting a translation on ψ, we can suppose that $\psi(X)$ contains the identity element e of C. Let A' be the subgroup of C generated by $\psi(X)$; as X is complete, the same is true of $\psi(X)$, thus also of A'. The group $A' \cap B$ being both complete and linear is necessarily finite. Because it is the kernel of the projection $A' \to A$, the fact that φ factors as $X \to A' \to A$ shows that $A' \to A$ is an isomorphism (this is the definition of maximal maps). The extension C is thus trivial, which proves the proposition. $\square$

We return now to the proof of theorem 9. To show that φ^* is *surjective*, we are going to use the extensions of J furnished by generalized Jacobians.

More precisely, let $P \in X$ and put $\mathfrak{m} = 2P$. The generalized Jacobian $J_\mathfrak{m}$ is an extension of J by a local group $L_\mathfrak{m}$. If t is a local uniformiser at P, the group $L_\mathfrak{m} = U_P^{(1)}/U_P^{(2)}$ is the group of functions of the form $1 + at + \cdots$ modulo those of the form $1 + bt^2 + \cdots$; it is thus the group $\mathbf{G}_a$. Let j_P be the element of $\text{Ext}(J, \mathbf{G}_a) = H^1(J, \mathcal{O}_J)$ corresponding to $J_\mathfrak{m}$ and put $j'_P = \varphi^*(j_P)$. We propose to determine j'_P. For this, we identify $H^1(X, \mathcal{O}_X)$ with the space $R/(R(0) + k(X))$ of classes of répartitions on X (cf. chap. II, prop. 3). With this identification we have:

Proposition 15. *The element j'_P corresponding to $J_\mathfrak{m}$ is equal to the class of the répartition defined by $r_P = 1/t$ and $r_Q = 0$ if $Q \neq P$.*

PROOF. Let U_i be an open cover of J such that there exist regular sections s_i of $J_\mathfrak{m}$ over U_i and suppose that $\varphi(P) \in U_0$. The $f_{ij} = s_j - s_i$ form a 1-cocycle with values in $\mathcal{O}_J$ and the class of this cocycle is j_P (cf. no. 5). The functions $f_{ij} \circ \varphi = g_{ij}$ thus define a 1-cocycle on X whose class is j'_P. Let $\varphi_\mathfrak{m}$ be the canonical map from X to $J_\mathfrak{m}$ normalized so that the composition $X \to J_\mathfrak{m} \to J$ is equal to φ. Putting $h_i = \varphi_\mathfrak{m} - \varphi \circ s_i$, the h_i thus define an element $h \in R/R(0)$ whose class is equal to $-j'_P$ and everything comes down to showing that h and $-r$ are in the same class. At every point $Q \neq P$, $h_Q = 0$ since $\varphi_\mathfrak{m}$ is regular at Q (chap. V, prop. 4). Thus it remains to show that $h_P = -1/t$, that is to say that *the polar part of h_0 at P is equal to $-1/t$.* This can be checked by using local symbols and their explicit determination in the case of the group $\mathbf{G}_a$ (cf. chap. III,

no. 3). The following method, due to Rosenlicht [68], has the advantage of
being applicable to arbitrary generalized Jacobians (cf. no. 20):

For every element λ of the projective line Λ, put $H_\lambda = t^{-1}(\lambda)$; if $Q \in X$,
put

$$D_Q = H_{t(Q)} - H_\infty = (t - t(Q)) = (1 - t/t(Q)).$$

The divisor $H_{t(Q)}$ can be written $Q + H'_Q$, where H'_Q is an effective divisor.
Furthermore, the map $Q \to H'_Q$ is a regular map from X to its symmetric
product, as we see by applying lemma 14 of chap. V to the covering t :
$X \to \Lambda$. Define a rational map $\psi : X \to J_\mathfrak{m}$ by the formula

$$\psi(Q) = \varphi_\mathfrak{m}(D_Q) = \varphi_\mathfrak{m}(Q) + \varphi_\mathfrak{m}(H'_Q) - \varphi_\mathfrak{m}(H_\infty). \qquad (36)$$

This map is regular away from the divisor of zeros of t. As D_Q is the divisor
of the function $1 - t/t(Q)$, $\psi(Q)$ is nothing other than the canonical image
of this function in the local group $L_\mathfrak{m} = \mathbf{G}_a$, that is to say $-1/t(Q)$. On
the other hand, when $Q = P$, the divisor H'_Q is prime to P since t has a
simple zero at P. Formula (36) then shows that $\psi(Q) - \varphi_\mathfrak{m}(Q)$ is a function
of Q which is regular at P. As the same is true of $\varphi_\mathfrak{m} - h_0 = \varphi \circ s_0$ (see
above), we conclude that $\psi - h_0$ is regular at P, that is to say that h_0 has
polar part $-1/t$, which completes the proof of the proposition.   □

We can now show that $\varphi^* : H^1(J, \mathcal{O}_J) \to H^1(X, \mathcal{O}_X)$ is surjective. Indeed, let ω be an element of the dual of $H^1(X, \mathcal{O}_X)$ which is orthogonal
to the image of φ^*. According to the duality theorem (chap. II, no. 8) ω
is identified with an everywhere regular *differential form* on X. If P is
any point of X and if t is a local uniformizer at P, proposition 15 shows
that the image of φ^* contains the répartition r equal to $1/t$ at P and to 0
elsewhere. Thus $\langle \omega, r \rangle = 0$ i.e., $\mathrm{Res}_P(\frac{1}{t}\omega) = 0$, which means that ω is zero
at P. This being the case for all P, we have $\omega = 0$, which shows that φ^* is
surjective and completes the proof of theorem 9.   □

Remark.   One can formulate proposition 15 more strikingly in terms of
tangent vectors. Let $\vec{v}$ be a tangent vector at the point P, with $\vec{v} \neq 0$. One
associates to $\vec{v}$ an element of $H^1(X, \mathcal{O}_X)$ in two different ways:

a) The vector $\vec{v}$ defines a linear form on $H^0(X, \underline{\Omega}^1)$, thus an element of its
dual $H^1(X, \mathcal{O}_X)$.

b) The vector $\vec{v}$ defines an isomorphism from the local group $L_\mathfrak{m}$ (with
$\mathfrak{m} = 2P$) to $\mathbf{G}_a$, and the Jacobian defines, thanks to this isomorphism, an
element of $H^1(J, \mathcal{O}_J)$ to which one then applies φ^*.

Proposition 15 then comes down to saying that the two elements of
$H^1(X, \mathcal{O}_X)$ thus defined *coincide*.

20. Polar part of the maps $\varphi_{\mathfrak{m}}$

Let $\mathfrak{m} = \sum n_P P$ be a modulus on X supported on S and let $\varphi_{\mathfrak{m}} : X \to J_{\mathfrak{m}}$ be the canonical map from X to the corresponding generalized Jacobian. The group $J_{\mathfrak{m}}$ is an extension of the usual Jacobian J by a local group $L_{\mathfrak{m}}$. By pull-back, $J_{\mathfrak{m}}$ defines a principal fiber space $P_{\mathfrak{m}}$ with base X and structural group $L_{\mathfrak{m}}$, and $\varphi_{\mathfrak{m}}$ defines a rational section of $P_{\mathfrak{m}}$. From this it follows, as in the preceding no., that to determine the fiber space $P_{\mathfrak{m}}$ comes down to determining the "polar part" of $\varphi_{\mathfrak{m}}$ at a point $P \in S$, that is to say to constructing a rational map $\psi : X \to L_{\mathfrak{m}}$ such that $\varphi_{\mathfrak{m}} - \psi$ is regular at P.

We have $L_{\mathfrak{m}} = (\prod U_P/U_P^{(n_P)})/\mathbf{G}_{\mathfrak{m}}$, and thus everything comes down to constructing a rational map $\psi_P : X \to U_P/U_P^{(n_P)}$ for a given point P. This construction is made in the following manner:

Let Δ be the diagonal of $X \times X$ and let F be a rational function on $X \times X$ whose divisor is of the form

$$(F) = \Delta + R \qquad \text{with} \quad (P, P) \notin \text{Supp}(R).$$

Such a function exists: if t is a local uniformizer at P, one can take $F(Q, Q') = t(Q) - t(Q')$. For every point Q of a non-empty open U of X, the partial function $F_Q(Q') = F(Q, Q')$ is a rational function of Q' which belongs to U_P and its class $\overline{F}_Q$ in the local group $U_P/U_P^{(n_P)}$ is well defined. One easily checks that $Q \to \overline{F}_Q$ is a regular map of U to the group $U_P/U_P^{(n_P)}$ and the argument of proposition 15 shows that *it is the desired map* ψ_P. Note the analogy between this definition and that of differentials on a curve given by Weil ([88], §II).

We also point out that, according to prop. 10 of chap. V, the local symbol $(\psi_P, g)_P$, $g \in U_P$, is equal to the inverse of $\overline{g} \in U_P/U_P^{(n_P)}$.

21. Cohomology of Abelian varieties

Let A be an Abelian variety of dimension g and let

$$H^*(A) = \sum_{n=0}^{\infty} H^n(A, \mathcal{O}_A)$$

be its cohomology algebra (cf. no. 17). The composition law

$$s : A \times A \to A$$

defines by passage to cohomology a homomorphism

$$s^* : H^*(A) \to H^*(A \times A).$$

According to the Künneth formula (proposition 12), the algebra $H^*(A \times A)$ is identified with $H^*(A) \otimes H^*(A)$. Using the fact that A has an identity element, one checks that for every $x \in H^n(A)$, $n > 0$,

$$s^*(x) = x \otimes 1 + \sum y_i \otimes z_i + 1 \otimes x, \qquad \deg(y_i) > 0, \quad \deg(z_i) > 0. \quad (37)$$

This identity means that the algebra $H^*(A)$, endowed with the map

$$s^* : H^*(A) \to H^*(A) \otimes H^*(A)$$

is a *Hopf algebra* in the sense of Borel [8], §6. But we have the following result:

Proposition 16. *Let H be an associative, anticommutative, connected (i.e., reduced to scalars in dimension 0) Hopf algebra. Let g be an integer such that $H^n = 0$ for $n > g$. Then* dim $H^1 \leq g$ *and, if equality holds, the algebra H is identified with the exterior algebra of the vector space H^1.*

PROOF. According to the structure theorem of Hopf-Borel ([8], theorem 6.1), the algebra H is the tensor product over the base field k of simple (i.e., generated by one element) algebras $k[x_i]$; put $n_i = \deg(x_i)$. The product of all the x_i is a non-zero element of H of degree equal to $\sum n_i$, whence the inequality $\sum n_i \leq g$. In particular, the number of x_i of degree 1 is $\leq g$; as this number is equal to dim H^1 we indeed have dim $H^1 \leq g$. If dim $H^1 = g$, all the x_i are necessarily of degree 1. Furthermore, their squares are all zero, for, if for example $x_1^2 \neq 0$, the product $x_1^2 \otimes x_2 \otimes \cdots \otimes x_g$ would be a non-zero element of H of degree $g + 1$, which is impossible. The algebra H is thus identified with the exterior algebra of H^1. □

We return now to the algebra $H^*(A)$:

Theorem 10. *If A is an Abelian variety of dimension g,* dim $H^1(A) = g$ *and the algebra $H^*(A)$ is identified with the exterior algebra of $H^1(A)$.*

PROOF. We first observe that $H^*(A)$ satisfies the hypotheses of proposition 16: $H^0(A) = k$ since A is complete and connected and $H^n(A) = 0$ for $n > g$ since A has dimension g ([76], thm. 2, or Grothendieck [27], theorem 3.6.5). Thus dim $H^1(A) \leq g$ and the theorem will be proved when we have proved the opposite inequality dim $H^1(A) \geq g$.

According to the corollary to theorem 4 (no. 13), there exists a strictly exact sequence

$$0 \to B \to C \to A \to 0 \quad (38)$$

where C is a product of Jacobians and where B is an Abelian variety. As $\text{Hom}(B, \mathbf{G}_a) = 0$, the exact sequence of Ext associated to (38) is

$$0 \to \text{Ext}(A, \mathbf{G}_a) \to \text{Ext}(C, \mathbf{G}_a) \to \text{Ext}(B, \mathbf{G}_a).$$

Taking into account theorem 7, we get the exact sequence

$$0 \to H^1(A) \to H^1(C) \to H^1(B),$$

whence the inequality

$$\dim \ H^1(A) \geq \dim \ H^1(C) - \dim \ H^1(B). \tag{39}$$

According to theorem 9, $\dim \ H^1(J) = \dim \ J$ for all Jacobians J; by virtue of the Künneth formula, the same is true of C which is a product of Jacobians. On the other hand, $\dim \ H^1(B) \leq \dim \ B$, as we have seen. The inequality (39) thus can be written

$$\dim \ H^1(A) \geq \dim \ C - \dim \ B = \dim \ A,$$

which completes the proof according to what has been said before.   □

Remark. Let $\underline{\Omega}^r$ be the sheaf of regular differential forms of degree r on A and put $h^{r,s} = \dim \ H^s(A, \underline{\Omega}^r)$. From the fact that the tangent bundle to a group variety is trivial (chap. III, prop. 16), the sheaf $\underline{\Omega}^r$ is isomorphic to the direct sum of $\binom{g}{r}$ copies of the sheaf $\mathcal{O}_A$ and theorem 10 gives

$$h^{r,s} = \binom{g}{r} \cdot \binom{g}{s}.$$

In particular the *symmetry formula* $h^{r,s} = h^{s,r}$ is valid for Abelian varieties; one knows that, in characteristc $p > 0$, there are non-singular varieties which violate this formula, cf. [77].

22. Absence of homological torsion on Abelian varieties

We suppose that the characteristic of the base field is $p > 0$. One knows ([77], §1) that one can associate to every algebraic variety X *Bockstein operations* acting on $H^*(X)$. One says that X has no *homological torsion* if these operations are identically zero.

Theorem 11. *An Abelian variety has no homological torsion.*

PROOF. Let A be an Abelian variety and let $\beta_1, \ldots, \beta_n, \ldots$ be the Bockstein operations associated to A. We assume that $\beta_i = 0$ for $i < n$ and show that $\beta_n = 0$. As β_n acts on the cohomology algebra of β_{n-1} (*loc. cit.* no. 3), we see that β_n acts on $H^*(A)$. Furthermore, β_n satisfies the following derivation formula (*loc. cit.*, formula (8)):

$$\beta_n(x.y) = \beta_n(x).F^n(y) + (-1)^{\deg(x)} F^n(x)\beta_n(y), \quad x, \ y \in H^*(A)$$

where F denotes the endomorphism of $H^*(A)$ defined by the p-th power on $\mathcal{O}_A$.

Since, by virtue of theorem 10, $H^*(A)$ is generated by its elements of degree 1, it will suffice to show that $\beta_n(x) = 0$ if $x \in H^1$. In this case, evidently $s^*(x) = x \otimes 1 + 1 \otimes x$, in other words that x is a *primitive* element of $H^1(A)$ (cf. no. 17). The functorial character of β_n then shows that

$y = \beta_n(x)$ is a primitive element of degree 2. But the fact that $H^*(A)$ is an exterior algebra implies, as is easily seen, that every non-zero primitive element of $H^*(A)$ is of degree 1; thus $y = 0$, as was to be shown. $\qquad\square$

Corollary 1. *Let $\varphi : X \to A$ be an everywhere regular maximal map from a complete variety X to an Abelian variety A. If Z'_∞ denotes the intersection of the kernel of the Bockstein operations β_n acting on $H^1(X, \mathcal{O}_X)$ (cf. [77], no. 7), then* dim $A \le$ dim Z'_∞.

PROOF. Proposition 14 shows that $\varphi^* : H^1(A, \mathcal{O}_A) \to H^1(X, \mathcal{O}_X)$ is injective. Since the β_n are zero on $H^1(A, \mathcal{O}_A)$, the image of φ^* is contained in Z'_∞. As dim $A =$ dim $H^1(A, \mathcal{O}_A)$ according to theorem 10, we deduce the desired inequality. $\qquad\square$

Remark. In fact, the corollary above gives an *exact bound* for dim A, in other words the dimension of the Albanese variety of X is *equal* to dim Z'_∞, cf. Mumford [**117**], p. 196.

Corollary 2. *Let $\varphi : X \to J$ be the canonical map of a non-singular projective curve to its Jacobian. For every connected unipotent group B, the homomorphism*

$$\varphi^* : H^1(J, B_J) \to H^1(X, B_X)$$

is bijective.

PROOF. We know that φ^* is injective (proposition 14) and that it is bijective if $B = G_a$ (theorem 9). We are going to pass from this to the general case by induction on dim B. If dim $B \ge 2$, there exists a strictly exact sequence

$$0 \to B' \to B \to B'' \to 0$$

where B' and B'' are connected unipotent groups of dimension strictly less than that of B. From this we deduce a commutative diagram

$$
\begin{array}{ccccccc}
0 & \longrightarrow & H^1(J, B'_J) & \longrightarrow & H^1(J, B_J) & \longrightarrow & H^1(J, B''_J) \\
& & \downarrow & & \downarrow & & \downarrow \\
0 & \longrightarrow & H^1(X, B'_X) & \longrightarrow & H^1(X, B_X) & \longrightarrow & H^1(X, B''_X)
\end{array}
$$

where the two end vertical arrows are bijections, in view of the induction hypothesis. If one knew that the homomorphism $H^1(J, B_J) \to H^1(J, B''_J)$ were surjective, it would follow that the middle vertical arrow was a surjection, which would establish the corollary; as J has no homological torsion, we are reduced to the following lemma:

Lemma 8. *Let $0 \to B' \to B \to B'' \to 0$ be a strictly exact sequence of unipotent groups, the group B being connected. If a variety X has*

no homological torsion in dimension q, the homomorphism $H^q(X, \mathcal{B}_X) \rightarrow$
$H^q(X, \mathcal{B}''_X)$ is surjective.

PROOF. The group B'' is connected; we argue by induction on its dimension. If dim $B'' = 0$, there is nothing to prove. If dim $B'' = 1$, then
$B'' = \mathbf{G}_a$. According to theorem 3, the group B is the quotient of a product
W of Witt groups, and it suffices to show that $H^q(X, \mathcal{W}_X) \rightarrow H^q(X, \mathcal{O}_X)$
is surjective. If $W = \prod W_{n_i}$, at least one of the homomorphisms $W_{n_i} \rightarrow \mathbf{G}_a$
has a non-trivial tangent map, which reduces us to the case of a surjective
separable homomorphism $f : W_n \rightarrow \mathbf{G}_a$. Such a homomorphism factors as
$W_n \xrightarrow{\varphi} \mathbf{G}_a \xrightarrow{g} \mathbf{G}_a$ with $\varphi = R^{n-1}$ (cf. no. 8), the homomorphism g being
separable. This means that

$$g(t) = a_0 t + a_1 t^p + \cdots + a_k t^{p^k}, \qquad \text{with } a_0 \neq 0.$$

Denoting by F the endomorphism of $V = H^q(X, \mathcal{O}_X)$ given by the p-th
power, we see that the endomorphism g_* of V defined by g is equal to

$$a_0 + a_1 F + \cdots + a_k F^k.$$

As V is a finite-dimensional vector space and F is p-linear, we conclude
that g_* is surjective (its differential is surjective, cf. chap. VI, no. 4). On
the other hand, φ_* is surjective since X has no homological torsion in
dimension q (cf. [77], no. 3); the homomorphism $f_* = g_* \circ \varphi_*$ is thus also
surjective, which proves the lemma when dim $B'' = 1$.

Now suppose that dim $B'' \geq 2$ and let

$$0 \rightarrow C'' \rightarrow B'' \rightarrow D'' \rightarrow 0$$

be a strictly exact sequence, where C'' and D'' are connected and of dimension $<$ dim B''. Let C be the inverse image of C'' in B and let C_0 be the
connected component of the identity element of C. In order to simplify the
notation put $T(B) = H^q(X, \mathcal{B}_X)$, and similarly for B'', C'', D'', and C_0.
If b'' is an element of $T(B'')$, the induction hypothesis applied to $B \rightarrow D''$
shows that there exists $b \in T(B)$ having the same image as b'' in $T(D'')$;
by subtraction, we are thus reduced to the case of an element $b'' \in T(B'')$
which gives 0 in $T(D'')$. The exact sequence

$$T(C'') \rightarrow T(B'') \rightarrow T(D'')$$

then shows that there exists $c'' \in T(C'')$ having image b''; the induction
hypothesis applied to $C_0 \rightarrow C''$ permits us to lift c'' to $c_0 \in T(C_0)$, and,
as $T(C_0)$ maps to $T(B)$, we finally do find an element of $T(B)$ with image
b'', as was to be shown.   □

23. Application to the functor $\mathrm{Ext}(A, B)$

Theorem 12. *If A is an Abelian variety, the functor $\mathrm{Ext}(A, B)$ is an exact functor on the category of (commutative) linear groups.*

PROOF. Put $T(B) = \mathrm{Ext}(A, B)$. We are going to show that, if we have a strictly exact sequence of linear groups

$$0 \to B' \to B \to B'' \to 0,$$

the corresponding sequence

$$0 \to T(B') \to T(B) \to T(B'') \to 0$$

is exact. In view of proposition 3, it suffices to prove that $T(B) \to T(B'')$ is surjective.

First we are going to treat several particular cases:

a) *B is a finite group*. One knows (Weil [89], p. 128) that, for every integer n, the map $x \to nx$ is an isogeny of A to itself. Choosing n to be a multiple of the order of G, we easily deduce that $T(G) = \mathrm{Ext}(A, G)$ is identified with $\mathrm{Hom}(_nA, G)$, denoting by $_nA$ the subgroup of A formed by elements x such that $nx = 0$. After decomposing G into a direct sum, we can also suppose that n is a power of a prime number. The group $_nA$ is then a direct sum of a certain number of cyclic groups of order n (Weil, loc. cit.), and the homomorphism $\mathrm{Hom}(_nA, B) \to \mathrm{Hom}(_nA, B'')$ is indeed surjective.

b) *B'' is a finite group*. Let B'_0 be the connected component of the identity element of B'. If the characteristic of k is zero, the group B is the product of B'_0 by the finite group B/B'_0, and we are reduced to a). If the characteristic of k is non-zero, we begin by removing the factors of type $\mathbf{G}_m$ from B'_0 (they are direct factors in B). Having done this, the group B is of finite period, and, lifting to B generators of B'', we see that there exists a finite subgroup C of B projecting onto B''; we then apply a) to $C \to B''$.

c) *B is a torus*. The same is then true of B''; thus $B = (\mathbf{G}_m)^r$ and $B' = (\mathbf{G}_m)^s$. The homomorphism $\varphi : B \to B''$ is defined by a matrix Φ with integral coefficients. As φ is surjective, there exists a matrix Ψ with integral coefficients such that $\Phi.\Psi = N$, where N is a non-zero integer. We have $T(B) = \mathrm{Ext}(A, \mathbf{G}_m)^r = P(A)^r$, denoting by $P(A)$ the dual variety of A (cf. no. 16); similarly, $T(B'') = P(A)^s$. The matrices Φ and Ψ define homomorphisms $P(\Phi)$ and $P(\Psi)$ satisfying $P(\Phi).P(\Psi) = N$. From the fact that $P(A)$ is an Abelian variety, multiplication by N is surjective and the same is true of $P(\Phi)$, which proves the desired result.

d) *B is unipotent and connected*. According to theorem 8, $H^1(A, \mathcal{B}_A) = T(B)$ and lemma 8 shows that $T(B) \to T(B'')$ is surjective.

e) *B is connected*. One decomposes B and B'' into a product of a torus and a unipotent group and applies c) and d).

Now we pass to the general case. Let B_0'' be the connected component of the identity in B'' and let B_0 be its inverse image in B. Let $b'' \in T(B'')$ and let x_0'' be the image of b'' in $T(B''/B_0'') = T(B/B_0)$. Applying b) to $B \to B/B_0$, we see that x_0'' is the image of an element of $T(B)$; by subtraction, we are reduced to the case where $x_0'' = 0$. The element b'' then comes from an element $b_0'' \in T(B_0'')$. If B_1 denotes the connected component of the identity element in B, we can apply e) to $B_1 \to B_0''$ and there exists $b_1 \in T(B_1)$ having image b_0''. As $T(B_1)$ maps to $T(B)$, we finally get an element of $T(B)$ with image b'', which finishes the proof. $\square$

One can give other cases where the functor $\mathrm{Ext}(A, B)$ is exact. We limit ourselves to the following:

Theorem 13. *Let C be an extension of an Abelian variety by a (commutative) connected linear group L. If G is a finite group, there is an exact sequence*

$$0 \to \mathrm{Ext}(A, G) \to \mathrm{Ext}(C, G) \to \mathrm{Ext}(L, G) \to 0.$$

PROOF. In view of proposition 2 it suffices to show that $\mathrm{Ext}(C, G) \to \mathrm{Ext}(L, G)$ is surjective, that is to say that every isogeny of L "extends" to C. Thus let $L' \in \mathrm{Ext}(L, G)$. According to theorem 12, the homomorphism $\mathrm{Ext}(A, L') \to \mathrm{Ext}(A, L)$ is surjective; there thus exists $C' \in \mathrm{Ext}(A, L')$ having image $C \in \mathrm{Ext}(A, L)$. The group C' contains L' as a subgroup, which contains G; the group C'/G is identified with C. One can thus consider C' as an element of $\mathrm{Ext}(C, G)$ and it is clear that this element has image L' in $\mathrm{Ext}(L, G)$, as was to be shown. $\square$

Example. We take for C a *generalized Jacobian* $J_\mathfrak{m}$, the Abelian variety A then being the usual Jacobian and the group L being the local group $L_\mathfrak{m}$ (chap. V, §3). We know (chap. VI, no. 12) that the group $\mathrm{Ext}(J_\mathfrak{m}, G)$ is identified with the group of classes of coverings of the curve having Galois group G and whose conductor is $\leq \mathfrak{m}$; similarly, the group $\mathrm{Ext}(J, G)$ is identified with the subgroup of classes of unramified coverings. Theorem 13 then shows that the quotient group is identified with $\mathrm{Ext}(L_\mathfrak{m}, G)$, a group whose definition is *purely local*.

Bibliographic note

Extensions of an Abelian variety A by the group $\mathbf{G}_a$ or the group $\mathbf{G}_m$ appeared for the first time in a short note of Weil [90]. This note contains the fact that $\mathrm{Ext}(A, \mathbf{G}_m)$ is isomorphic to the group of classes of divisors X on A such that $X \equiv 0$.

This result is recovered by Barsotti [4], who systematically takes the point of view of "factor systems". Barsotti also determines the dimension of $\mathrm{Ext}(A, \mathbf{G}_a)$, thanks to the classification of purely inseparable isogenies.

The relation $H^1(A, \mathcal{O}_A) = \mathrm{Ext}(A, \mathbf{G}_a)$ is proved by Rosenlicht [68] (see also Barsotti [6], as well as [78]), and he obtains the dimension of $H^1(A, \mathcal{O}_A)$ by means of generalized Jacobians. It is his proof that we have given, with a few variations.

Recently Cartier has obtained a result more precise than this simple dimension computation: he has established a "functorial" isomorphism between $H^1(A, \mathcal{O}_A)$ and the tangent space t_{A^*} of the dual variety A^* of A, and from this he deduces the "biduality theorem" $A^{**} = A$, cf. [13], [107].

Finally, the fact that every connected commutative unipotent group is isogenous to a product of Witt groups was proved by Chevalley and Chow (non-published), as well as Barsotti [7]. According to Dieudonné [23], an analogous result holds in "formal" geometry. See Cartier [108] and Demazure-Gabriel [112].

Bibliography

1. E. Artin and H. Hasse, *Die beiden Ergänzungssätze zum Reziprozitäts-gesetz der l^n-ten Potenzreste im Körper der l^n-ten Einheitswurzeln*, Hamb. Abh. **6** (1928), 146-162.

2. R. Baer, *Erweiterungen von Gruppen und ihren Isomorphismen*, Math. Zeit. **38** (1934), 375-416.

3. I. Barsotti, *A note on abelian varieties*, Rend. Cir. Palermo **2** (1954), 1-22.

4. I. Barsotti, *Structure theorems for group varieties*, Annali di Mat. **38** (1955), 77-119.

5. I. Barsotti, *Abelian varieties over fields of positive characteristic*, Rend. Cir. Palermo **5** (1956), 1-25.

6. I. Barsotti, *Repartitions on abelian varieties*, Illinois J. of Maths. **2** (1958), 43-70.

7. I. Barsotti, *On Witt vectors and periodic group-varieties*, Illinois J. of Maths. **2** (1958), 99-110 and 608-610.

8. A. Borel, *Sur la cohomologie des espaces fibrés principaux et des espaces homogènes de groupes de Lie compacts*, Ann. of Maths. **57** (1953), 115-207.

9. A. Borel, *Groupes linéaires algébriques*, Ann. of Maths. **64** (1956), 20-82.

10. H. Cartan and S. Eilenberg, "Homological algebra," Princeton Math. Ser., no. 19.

11. H. Cartan and C. Chevalley, *Géometrie algébrique*, Séminaire E.N.S. (1955-1956).

12. P. Cartier, *Une nouvelle opération sur les formes différentielles*, Comptes Rendus **244** (1957), 426-428.

13. P. Cartier, *Dualité des variétés abéliennes*, Séminaire Bourbaki (May, 1958).

14. C. Châtelet, *Variations sur un thème de H. Poincaré*, Annales E.N.S. **61** (1944), 249-300.

15. C. Chevalley, "Introduction to the theory of algebraic functions of one variable," Math. Surv., VI, New York, 1951.

16. C. Chevalley, "Class field theory," Nagoya, 1954.

17. C. Chevalley, *Classification des groupes de Lie algébriques*, Séminaire E.N.S. (1956-58).

18. W. L. Chow, *The Jacobian variety of an algebraic curve*, Amer. J. of Maths. **76** (1954), 453-476.

19. W. L. Chow, *On the projective embedding of homogeneous varieties*, Symp. in honor of S. Lefschetz (1957), 122-128.

20. A. Clebsch and P. Gordan, "Theorie der abelschen Funktionen," Teubner, Leipzig, 1866.

21. M. Deuring, *Die Zetafunktion einer algebraischen Kurve vom Geschlechte eins (vierte Mitteilung)*, Gött. Nach. (1957), no. 3.

22. J. Dieudonné, "La géométrie des groupes classiques," Ergeb. der Math., no. 5, Springer, 1955.

23. J. Dieudonné, *Groupes de Lie et hyperalgèbres de Lie sur un corps de caractéristique $p > 0$ (VII)*, Math. Ann. **134** (1957), 114-133.

24. J. Dieudonné, *On the Artin-Hasse exponential series*, Proc. Amer. Math. Soc. **8** (1957), 210-214.

25. R. Godement, "Topologie algébrique et théorie des faisceaux," Hermann, Paris, 1958.

26. D. Gorenstein, *An arithmetic theory of adjoint plane curves*, Trans. Amer. Math. Soc. **72** (1952), 414-436.

27. A. Grothendieck, *Sur quelques points d'algèbre homologique*, Tohoku Math. J. **9** (1957), 119-221.

28. A. Grothendieck, *Théorèmes de dualité pour les faisceaux algébriques cohérents*, Séminaire Bourbaki (May, 1957).

29. A. Grothendieck, *Le théorème de Riemann-Roch (rédigé par A. Borel et J.-P. Serre)*, Bull. Soc. Math. France **86** (1958), 97-136.

30. H. Hasse, *Bericht über neuere Untersuchungen und Probleme aus der Theorie der algebraischen Zahlkörper*, Jahr. der D. Math. Ver. **35** (1926), 1-55; *ibid* **36** (1927), 255-311; *ibid* **39** (1930), 1-204.

31. H. Hasse, *Theorie der relativ-zyklischen algebraischen Funktionenkörper, insbesondere bei endlichem Konstantenkörper*, J. Crelle **172** (1934), 37-54.

32. H. Hasse, *Theorie der Differentiale in algebraischen Funktionenkörpern mit vollkommenem Konstantenkörper*, J. Crelle **172** (1934), 55-64.

33. J. Herbrand, "Le développement moderne de la théorie des corps algébriques—corps de classes et lois de réciprocité," Mém. des Sc. Maths., no. 75, Paris, 1936.

34. H. Hironaka, *On the arithmetic genera and the effective genera of algebraic curves*, Mem. Kyoto **30** (1957), 177-195.

35. F. Hirzebruch, "Neue topologische Methoden in der algebraischen Geometrie," Ergeb. der Math., no. 9, Springer, 1956.

36. G. Hochschild and T. Nakayama, *Cohomology in class field theory*, Ann. of Maths. **55** (1952), 348-366.

37. J. Igusa, *On some problems in abstract algebraic geometry*, Proc. Nat. Acad. Sci. USA **41** (1955), 964-967.

38. J. Igusa, *Fibre systems of Jacobian varieties*, Amer. J. of Maths. **78** (1956), 171-199.

39. J. Igusa, *Fibre systems of Jacobian varieties, II*, Amer. J. of Maths. **78** (1956), 745-760.

40. Y. Kawada, *Class formations, I*, Duke Math. J. **22** (1955), 165-178.

41. Y. Kawada, *Class formations, III*, J. Math. Soc. Japan **7** (1955), 453-490.

42. Y. Kawada, *Class formations, IV*, J. Math. Soc. Japan **9** (1957), 395-405.

43. Y. Kawada and I. Satake, *Class formations, II*, J. Fac. Sci., Tokyo **7** (1955), 353-389.

44. K. Kodaira, *On compact complex analytic surfaces I*, (mimeographed notes). Princeton (1955).

45. K. Kodaira, *The theorem of Riemann-Roch on compact analytic surfaces*, Amer. J. Maths. **73** (1951), 813-875.

46. E. Kolchin, *On certain concepts in the theory of algebraic matrix groups*, Ann. of Maths. **49** (1948), 774-789.

47. A. Krazer and W. Wirtinger, *Abelsche Funktionen und allgemeine Thetafunktionen*, Enc. Math. Wis., II B-7 (1920).

48. S. Lang, *Algebraic groups over finite fields*, Amer. J. of Maths. **78** (1956), 555-563.

49. S. Lang, *Unramified class field theory over function fields in several variables*, Ann. of Maths. **64** (1956), 285-325.

50. S. Lang, *Sur les séries L d'une variété algébrique*, Bull. Soc. Math. de France **84** (1956), 385-407.

51. S. Lang, "Introduction to algebraic geometry," Interscience Tracts no. 5, New York, 1958.

52. S. Lang, "Abelian varieties," Interscience Tracts no. 7, New York, 1959.

53. S. Lang and J.-P. Serre, *Sur les revêtements non ramifiés des variétés algébriques*, Amer. J. of Maths. **79** (1957), 319-330.

54. S. Lang and J. Tate, *Galois cohomology and principal homogeneous spaces*, Amer. J. of Maths. **80** (1958), 659-684.

55. M. Lazard, *Sur les groupes de Lie formels à un paramètre*, Bull. Soc. Math. de France **83** (1955), 251-274.

56. T. Matsusaka, *The theorem of Bertini on linear systems in modular fields*, Kyoto Math. Mem. **26** (1950), 51-62.

57. H. Morikawa, *Generalized Jacobian varieties and separable abelian extensions of function fields*, Nagoya Math. J. **12** (1957), 231-254.

58. M. Nagata, *On the embedding problem of abstract varieties in projective varieties*, Mem. Kyoto **30** (1956), 71-82.

59. D. Northcott, *The neighborhoods of a local ring*, Journ. London Math. Soc. **30** (1955), 360-375.

60. D. Northcott, *A note on the genus formula for plane curves*, Journ. London Math. Soc. **30** (1955), 376-382.

61. D. Northcott, *A general theory of one-dimensional local rings*, Proc. Glasgow Math. Ass. **2** (1956), 159-169.

62. I. Reiner, *Integral representations of cyclic groups of prime orders*, Proc. Amer. Math. Soc. **8** (1957), 142-146.

63. M. Rosenlicht, *Equivalence relations on algebraic curves*, Ann. of Math. **56** (1952), 169-191.

64. M. Rosenlicht, *Generalized Jacobian varieties*, Ann. of Math. **59** (1954), 505-530.

65. M. Rosenlicht, *A universal mapping property of generalized Jacobian varieties*, Ann. of Maths. **66** (1957), 80-88.

66. M. Rosenlicht, *Some basic theorems on algebraic groups*, Amer. J. of Maths. **78** (1956), 401-443.

67. M. Rosenlicht, *A note on derivations and differentials on algebraic varieties*, Port. Math. **16** (1957), 43-55.

68. M. Rosenlicht, *Extensions of vector groups by abelian varieties*, Amer. J. of Maths. **80** (1958), 685-714.

69. P. Samuel, *Singularités des variétés algébriques*, Bull. Soc. Math. de France **79** (1951), 121-129.

70. P. Samuel, "Algèbre locale," Mém. Sci. Math., no. 123, Paris, 1953.

71. P. Samuel, "Méthodes d'algèbre abstraite en géométrie algébrique," Ergeb. der Math., no. 4, Springer, 1955.

72. H. L. Schmid, *Über das Reziprozitätsgesetz in relativ-zyklischen algebraischen Funktionenkörpern mit endlichem Konstantenkörper*, Math. Zeit. **40** (1936), 94-109.

73. J.-P. Serre, *Faisceaux algébriques cohérents*, Ann. of Maths. **61** (1955), 197-278.

74. J.-P. Serre, *Cohomologie et géométrie algébrique*, Int. Cong. Amsterdam (1954, vol. III), 515-520.

75. J.-P. Serre, *Géométrie algébrique et géométrie analytique*, Annales Inst. Fourier **6** (1955-56), 1-42.

76. J.-P. Serre, *Sur la cohomologie des variétés algébriques*, J. de Maths. pures et appl. **36** (1957), 1-16.

77. J.-P. Serre, *Sur la topologie des variétés algébriques en caractéristique p*, Symposium de topologie algébrique, Mexico (1956), 24-53.

78. J.-P. Serre, *Quelques propriétés des variétés abéliennes en caractéristique p*, Amer. J. of Maths. **80** (1958), 715-739.

79. F. Severi, "Vorlesungen über algebraische Geometrie," Teubner, Leipzig, 1921.

80. F. Severi, "Serie, sistemi d'equivalenza e corrispondenze algebriche sulle varietà algebriche," Rome, 1942.

81. F. Severi, "Funzioni quasi abeliane," Vatican, 1947.

82. G. Shimura and Y. Taniyama, *Complex multiplication of abelian varieties and its applications to number theory*, J. Math. Soc. Japan 10 (1958).

83. J. Tate, *The higher dimensional cohomology groups of class field theory*, Ann. of Maths. 56 (1952), 294-297.

84. J. Tate, WC-*groups over p-adic fields*, Séminaire Bourbaki, Dec., 1957.

85. A. Weil, *Zur algebraischen Theorie der algebraischen Funktionen*, J. Crelle 179 (1938), 129-133.

86. A. Weil, *Sur les fonctions algébriques à corps de constantes finis*, Comptes Rendus 210 (1940), 592-594.

87. A. Weil, "Foundations of algebraic geometry," Coll. no. 29, New York, 1946.

88. A. Weil, "Sur les courbes algébriques et les variétés qui s'en déduisent," Hermann, Paris, 1948.

89. A. Weil, "Variétés abéliennes et courbes algébriques," Hermann, Paris, 1948.

90. A. Weil, "Variétés abéliennes," Colloque d'algébre et théorie des nombres, Paris, 1949, pp. 125-128.

91. A. Weil, *Sur la théorie du corps de classes*, J. Math. Soc. Japan 3 (1951), 1-35.

92. A. Weil, *Abstract versus classical algebraic geometry*, Int. Cong. Amsterdam, 1954, vol. III, 550-558.

93. A. Weil, *On algebraic groups of transformations*, Amer. J. of Maths. 77 (1955), 355-391.

94. A. Weil, *On algebraic groups and homogeneous spaces*, Amer. J. of Maths. 77 (1955), 493-512.

95. A. Weil, *The field of definition of a variety*, Amer. J. of Maths. 78 (1956), 509-524.

96. A. Weil, "Introduction à l'étude des variétés kählériennes," Hermann, Paris, 1958.

97. H. Weyl, "Die Idee der Riemannschen Fläche," Teubner, Leipzig, 1923.

98. G. Whaples, *Local theory of residues*, Duke Math. J. 18 (1951), 683-688.

99. E. Witt, *Zyklische Körper und Algebren der Charakteristik p vom Grade p^n*, J. Crelle 176 (1936), 126-140.

100. N. Yoneda, *On the homology theory of modules*, J. Fac. Sci. Tokyo 7 (1954), 193-227.

101. O. Zariski, *Pencils on an algebraic variety and a new proof of a theorem of Bertini*, Trans. Amer. Math. Soc. 50 (1941), 48-70.

102. O. Zariski, *Complete linear systems on normal varieties and a generalization of a lemma of Enriques-Severi*, Ann. of Maths. **55** (1952), 552-592.

103. O. Zariski, *Scientific report on the second summer institute. Part III. Algebraic sheaf theory.*, Bull. Amer. Math. Soc. **62** (1956), 117-141.

104. H. Zassenhaus, "The theory of groups," Chelsea, New York, 1949.

Supplementary Bibliography

It concerns the following questions:

— Class field theory: Artin-Tate [106]; Cassels-Fröhlich [109]; Lang [116]; Weil [125].
— Foundations of algebraic geometry: Grothendieck [113], [114].
— Duality, differential forms, residues: Altman-Kleiman [105]; Hartshorne [115]; Tate [124].
— Abelian varieties: Cartier [107]; Mumford [118]; Néron [119].
— Picard variety: Chevalley [110], [111]; Grothendieck [114]; Mumford [117]; Raynaud [121];
— Generalized Jacobians: Oort [120].
— Unipotent algebraic groups: Cartier [108], Demazure-Gabriel [112]; [122].
— Proalgebraic groups and applications to class field theory: [122], [123]; Hazewinkel [112].

105. A. Altman and S. Kleiman, "Introduction to Grothendieck Duality Theory," Lect. Notes in Math. no. 146, Springer, 1970.
106. E. Artin and J. Tate, "Class Field Theory," Benjamin, New York, 1967.
107. P. Cartier, *Isogenies and duality of abelian varieties*, Ann. of Math. 71 (1960), 315-351.
108. P. Cartier, *Groupes algébriques et groupes formels*, Colloque sur la théorie des groupes algébriques, Bruxelles (1962), 87-111.
109. J. Cassels and A. Fröhlich (eds.), "Algebraic Number Theory," Academic Press, New York, 1967.
110. C. Chevalley, *Sur la théorie de la variété de Picard*, Amer. J. of Math. 82 (1960), 435-490.

111. C. Chevalley, *Variétés de Picard*, Séminaire E.N.S., 1958-1959.
112. M. Demazure et P. Gabriel, "Groupes Algébriques I (with an appendix "Corps de classes local" by M. Hazewinkel)," Masson et North-Holland, Paris, 1970.
113. A. Grothendieck, *Eléments de Géométrie Algébrique* (rédigés avec la collaboration de J. Dieudonné). *Chap. 0 à IV*, Publ. Math. I.H.E.S. **4, 8, 11, 17, 20, 24, 28, 32** (1960-1967).
114. A. Grothendieck, *Fondements de la Géométrie Algébrique*, (extraits du Séminaire Bourbaki) (1962), Secr. Math., rue P.-Curie, Paris.
115. R. Hartshorne, "Residues and Duality," Lect. Notes in Math. no. 20, Springer, 1966.
116. S. Lang, "Algebraic Number Theory," Addison-Wesley, Reading, 1970.
117. D. Mumford, "Lectures on Curves on an Algebraic Surface," Ann. of Math. Studies, no. 59, Princeton, 1966.
118. D. Mumford, "Abelian Varieties," Oxford, 1970.
119. A. Néron, *Modèles minimaux des variétés abéliennes sur les corps locaux et globaux*, Publ. Math. I.H.E.S. **21** (1964), 3-128.
120. F. Oort, *A construction of generalized Jacobian varieties by group extensions*, Math. Ann. **147** (1962), 277-286.
121. M. Raynaud, *Spécialisation du foncteur de Picard*, Publ. Math. I.H.E.S. **38** (1970), 27-76.
122. J.-P. Serre, *Groupes proalgébriques*, Publ. Math. I.H.E.S. **7** (1960), 339-403.
123. J.-P. Serre, *Sur les corps locaux à corps résiduel algébriquement clos*, Bull. Soc. Math. de France **89** (1961), 105-154.
124. J. Tate, *Residues of differentials on curves*, Ann. Sci. E.N.S. **1** (4) (1968), 149-159.
125. A. Weil, "Basic Number Theory (3rd edition)," Springer, New York, 1974.

Index

Graduate Texts in Mathematics

continued from page ii